AF882106

INSIDE the Rise of Tech

Chronicles of a Tech Consultant

by
Eva Springstone

Copyright © 2020 Eva Springstone
All rights reserved

No part of this book may be reproduced, or stored in a retrieval system, or transmitted in any form or by any means, electronic, mechanical, photocopying, recording, or otherwise, without express written permission of the publisher.

Cover and interior design by the author

This is a work of creative non-fiction. All of the events in this memoir are true to the best of the author's memory. Some names and identifying features have been changed to protect the identity of certain parties. The author in no way represents any company, corporation, or brand, mentioned herein. The views expressed in this memoir are solely those of the author.

Throughout this book trademarks are used. Rather than put a trademark symbol in every occurrence of a trademarked name, I am using the names in an editorial fashion with no intention of infringement of the trademark. No such use, or the use of any trade name is intended to convey endorsement or other affiliation with this book.

ISBN 978-1-7349903-1-7

Contents

CH 1. What This Book Is About **1**
1.1 What's a Tech Consultant? 3
1.2 Is Tech for you? 5

CH 2. How My Foot Got in Tech's Door **7**
2.1 My first step 7
2.2 My second step 11
2.3 My final step 13

CH 3. Groupie To the Early Geeks **17**
3.1 They don't like to talk 18
3.2 Elegant code 22
3.3 Extreme perspicacity 26
3.4 High-speed innovation machine 28
3.5 A woman among men 36

CH 4. The Lords of Tech **41**
4.1 The vision of new Tech vs. the reality 43
4.2 Tricks of the trade 49
4.3 Users—Tech's beasts of burden 80

CH 5. www.whatjusthappened?.com 89
5.1 That crazy thing called HTML 93
5.2 The usability profession exits the race 96
5.3 UI Design: breaking new ground 99
5.4 Help! We need some user satisfaction! 102
5.5 From Search and Browse to Buy and Sell 105
5.6 Bubblemania: startup creativity vs. cookie-cutter code 109

CH 6. Can't Get No User Satisfaction? 113
6.1 Job Interview Part 1: the WHY of user satisfaction 114
6.2 Job Interview Part 2: the HOW of user satisfaction 119
6.3 Job Interview Part 3: the WHO of user satisfaction 138
6.4 Getting along with "them" 141

CH 7. My Lonely Masquerade 145
7.1 Developers are from Pluto, but where do UI Designers come from?. 148
7.2 Congratulations, you've been out-sourced! 151
7.3 Masquerade at a bank 161
7.4 Masquerades at pharma companies 164
7.5 Masquerades in government 173

CH 8. The Touchables 185
8.1 Touch me, babe! 188
8.2 Metamorphoses 197
8.3 The return to innocence 201

CH 9. Designing for Human Nature **203**
9.1 What humans want: A speculative exploration 204
9.2 What we want most from Tech 207
9.3 The never-ending story of Tech 222

AP 1. Career Building and Parenting **227**
Flexible Remote/In-Office option with Onsite Childcare 228
Nanny Services option 229
Regular Time-At-Home option 230
The Ethical Employer 230

AP 2. Business Analyst Job Description **233**

AP 3. Use Case Example **235**

AP 4. My Early UI Design Publications and Presentations (redacted) . **241**

AP 5. Masquerading Job Descriptions **243**

References **253**

About The Author **257**

CH 1. What This Book Is About

From 1989-2015, I was a consultant in the business software side of information technology (IT). I was there in the early days, when business software ran only on a Mainframe[1]. I took part in that early Tech and the rest that followed it, slam-jamming and slip-sliding along the Tech speedway over the next 25 years.

Yes, I rode that speedway of endless opportunity and excitement, but I wasn't a driver until the very end, when I challenged the rules and won. Before that, only those who controlled the code had driving privileges. They were the ones who either wrote the code or managed the code that was written. We called them "developers." They drove the Tech cars, and they didn't really like to take anyone else along except the Cleanup Crew. That's the crew I was a part of when I became a Tech Consultant in 1989.

In those early days of Tech, a software company's Cleanup Crew was minimal—a few S&M guys (sales and marketing), who sold business customers on the vision of the new software, and one or two technical writers, who wrote down the instructions for using that software. But when the software business started growing like crazy, the demands for cleanup on Tech projects steadily increased. Now, what is called the "Software Product Team," typically also includes business analysts, testers, hardware and software configurators and architects, designers, customer support, product "evangelists", and stakeholders. In my travels through Tech, these bloated teams were often a mix of Cleanup Crew, who did actual work, and hangers-on.

But even though it has increased in personnel, today's Cleanup Crew is still just a ride-along in the race cars of Tech. Why can't they drive? The members of the Cleanup Crew have a job only because their skills are needed to clean up after the *code* that is written. They don't write the code, and without code, there is nothing for them to do. It doesn't matter that Cleanup Crew jobs can be amazingly interesting and rewarding, and that you have to know a decent amount of Tech to do those jobs. The bottom line is Cleanup Crew members don't drive the Tech cars, and when times get tough, the cleaners are expendable. Those who join the Cleanup Crew know this but want to ride the speedway anyway, and here's why. Every now and then, they get into a Tech car that the developers are driving hot. They are driving it like they stole it, pedal to the metal. That's when the Cleanup Crew just might get to be a part of some Tech that ends up changing the world. And that's a helluva ride.

All in all, it's always been a great time for the Cleanup Crew, as long as you know your place and have the skill to leap gracefully out of one Tech car and into another—before you're kicked to the curb or the car crashes. Once I mastered that skill, I became a very successful Tech Consultant. I saw the rise of Tech from the inside, from mainframes to Alexa.

I wrote this book because I want to tell the story of how business software was developed done during my time, the good and the bad. I want to explain to people who were stuck with using crappy software what they were never told—why that shiny, new business software was so hard to use.

There's something else I want to tell, too. Sick of cleaning up after bad software, I decided to learn how to make software that people would like to use. That's when I became a User Experience (UX) Design Consultant, and—rules be damned—I finally broke free and drove my own Tech car. It wasn't easy, and in this book, I tell how getting to that point was the worst and then the best part of my time in Tech.

Part of my story describes how my success was intertwined with some of the exciting Tech that's coming down the speedway now. We are so lucky that some of it was made right, but it's not a mystery why. I explain why the right Tech happened and why people love using it. My hope is that, if we know that software can be made right, then we will settle for nothing less.

I finish this book with a discussion of where we need the new Tech to go and how to be part of it, in case you might be interested in a Tech consulting career.

If you are wondering what the life of a Tech consultant is like, here's a quick overview of how it was for me.

1.1 What's a Tech Consultant?

In Tech, there are contractors and consultants (or "conslutants", as the Tech world sometimes refers to us). I've been called both. From my perspective, the only difference is this: if I want to sound fancy, I'm a Consultant. In any case, the good ones, like me, know their stuff and exude reliability and confidence. The rewards of this career are great, whether you are part of a Cleanup Crew or a UX Designer, as my story shows.

Even though my race is mostly run, I can still successfully sell myself to top US banks, pharma companies, and Fortune 100 Tech companies. I go to the top bidder. That's the one that offers me the work I enjoy, for top dollar, in a location and environment that I like. I have no loyalty to the company. My loyalty is to the work. If the project turns out to be boring, or pointless, or I don't like the caliber of the people I work with, I usually move on. Recruiters from IT services companies (a.k.a. bodyshops) all over the country send me job prospects, and I know I can get just about any of those jobs whenever I'm ready to. As my story shows, it took a chunk of years and some hard lessons to get there, but at the top of my game, being a "consultant" was all I hoped it would be.

Of course, a consultant always comes into a company from the outside and always remains separate from the company employees. Is it hard to succeed in a job where you are an outsider? No, as long as you understand the consulting role you are playing and do high quality work. Based on my experience, here's the skinny on the atmosphere you face inside a company, once you've accepted a top consulting job or contract.

For me and the companies I choose to work for, it's typically a win-win situation. I work hard, turn in results that exceed expectations, and deliver on schedule, or earlier. I have fun and make great money. When I come into a project, I carry no gossip, no history, no friends inside and no enemies. I'm a fixer. One of my greatest assets to companies is that I can come in to a failing project with a dysfunctional team, take the reins, and drive my work to a successful outcome. My skills quickly command respect, and fairly soon it's clear that no one wants to stand in my way. Troublemakers get more resistance than they bargained for and typically give up or find another project to wreck. The rest don't bother me because they see smiles on the faces of management. When my work is completed successfully and the contract is up, the team's manager typically asks me to stay for more. But I have the glorious freedom to stay or go as the mood strikes me.

That's the typical scenario, and the times when things don't work out are few and far between. Whatever happens, I don't worry about having a job. There's always one out there for me. Even so, I do try to get along with my co-workers. Not only does this make the work more fun, it can also lead to future networking opportunities and friendships with like-minded colleagues.

Oddly enough, even though I often come in to take over the work that full-time employees (FTEs) have failed to complete, most of the time these same FTEs end up liking me and want to keep me around as long as possible. Why? They are part of the team I work with, and when I finish and turn in a stellar result, the company gives them the credit. Meh. I don't envy them. I know the manager will give me a great recommendation for my next job, and moving on to something new is what I care about. For many FTE's though, the primary goal is to save their current jobs, which are constantly threatened by the next merger or RIF (reduction in force, a.k.a., layoffs) that always seems to be lurking around the corner. It doesn't surprise me when I encounter an FTE who is afraid to go out on a limb. Doing so would risk failing or pissing someone off who might end up as your manager after the next merger.

Unlike consultants, FTEs often fear the process of updating their resumes. The possibility that they could be headed for the job market usually scares the crap out of them. To hold on to their job security, they don't take chances. As time passes, they become one-trick ponies, doing the same job every day. Many of them haven't learned new skills in years. They're not very marketable, and they know it, which makes resume updating even scarier. Rarely will they will take the ball and run with it. They typically fear innovation. I've seen FTE's cringe at the thought of challenging an established process, even though it could lead to better results. In my experience, most of them keep their heads down and don't ever chase career dreams.

I'm not like them, but I don't judge. Sometimes in life, you have to keep your head down and not rock the boat. Not to mention the fact that, as it turns out, I depend on them for my own consulting opportunities. Awash in their stultifying routines, the question arises that eventually leads to me. What can FTEs do when real work needs to be done? How can they handle challenges that require spirit, innovation, and the courage to stand up for unpopular ideas and methods? The answer is, they hire a consultant, and if they should be so lucky as to get one who delivers successful results and works well with the team, they have struck gold.

The beauty of it is that, because the consultant is temporary and expendable, he or she is never a threat to the FTE's position or status. Furthermore, company bean counters view consultants as cheap solutions, who improve their bottom line. Last but not least, because consultants are from outside, managers can give the company's internal awards to their FTEs for project successes. This makes managers look good and makes the FTEs feel more secure.

All in all, there's no downside for them or for me. Yeah, it's win-win.

1.2 Is Tech for you?

The Information Highway, The Age of Computers, the Technology Revolution, whatever you prefer to call it, is what changed the world and made it possible for me and zillions of other regular folk to chase their career dreams. I've been chasing my dreams for years. Academic professor was my first career. UX Design Consultant is my third, and likely my last. As a UX Designer, I'm committed to the people who have to use the software we make. I've seen the effects of software that ignores the needs of its users, and I won't abide it anymore. I know how bad software happens, I know how to make software that people love, and I've seen how doing so carries immense rewards. In this book, I'm going to tell the story of all of these things.

I dedicate this book to all who are disappointed spiritually, mentally, and monetarily with their jobs and think there are no career dreams left. Maybe my story will help them see that Tech is there for them, too, and it's a great ride—especially if they are interested in making software that people love.

With each Tech upheaval between 1989 and 2015—the mainframe, personal computing, the web, the iPhone, Alexa—the ground shifted under us. Our partitioned world of incommensurate cultures and limited career paths turned into an interconnected and expanding universe of opportunities, built upon the common language of Tech. Things got better, much better, but we've only just started the race.

Therefore, I also dedicate this book to the younger generation. Maybe my story will help them understand how Tech upheavals change the world and open the door to even more powerful, new Tech. In our time, this generation will be able to work with Tech that can understand our human nature and meet us on our own terms. They can unlock its power if they learn how to make software that people love and avoid the mistakes we made before.

CH 2. How My Foot Got in Tech's Door

Life is strange. I started a career in Tech out of necessity rather than choice. I initially thought of it as a temporary expedient. Now, looking back, that career seems so right for me, and I sometimes find it scary that it happened all by luck. On the other hand, I now see that my story, like many others, shows that we can make my own luck through determination, stubbornness, and the courage to try something new. In my case, I took 3 steps that made me lucky enough to get into the world of Tech.

2.1 My first step

In 1971, when I graduated from Northwestern University, the Data Center had only been around for a few years. That one-story building sat on the near side of the new lagoon, which was part of the recent campus renovation. Hustling to class by the lagoon, I allowed myself passing glances through the Data Center's floor to ceiling windows. I could see the huge mainframe machines with their quirky, 10-inch wide tape drives that jerked and stuttered back and forth as they shifted data around. Kind of cool but irrelevant to what was on my mind as a woman of the 60's. Across the lagoon to the east was the endless blue of Lake Michigan. A much more interesting view.

It was called the Data Center, not the Computing Center. What was it for? Computerized schedules and grade reports were the common beliefs. This was partially true, but the Data Center was also for the student nerds in the physics, engineering, and math classes who spent time there doing cardpunch[2] programming on a mainframe. Yes indeed. They used CARDS (3x8 cardstock paper) with rows of tiny black rectangles that were PUNCHED in various, meaningful sequences according to the programs they were writing. If you wanted to program a computer, or run a program through the computer, you had to be right there, right next to the huge machines. You had to create punched cards (the program) and then stack those cards ever-so-carefully in the correct order into the input bin so the monstrous mainframe could eat them. You needed a lot of cards to get the machine to do the simplest things. Every little punch meant something to the mainframe, and it manipulated, arranged, and created new data according to those punches. Then it spit your cards (your program) into the output bin. To run your program again, you had to put your cardstack back into the input bin in the original card order (of course, the cards were not numbered). Once a program was punched, its cardstack was preciously handled. A dropped cardstack scattered randomly on the floor was essentially an erased program.

But who ran the Data Center? It wasn't the department of Computer Science and its spawn--the developers, architects, code monkeys, hackers & geeks we know and love today. They didn't exist yet. We students of the 60's never thought much about who supported and maintained the Data Center's big machines or how those people contributed to the nascent techno-fabric of our lives. Most of us never went into the noisy building, just as we would never go into the facilities maintenance building where the janitors hung out. More to the point, we were preoccupied with important things to demonstrate about and smoke. Looking back, I can't say for sure who ran the Data Center, but if I had to guess now, I would say field support guys from IBM or individuals hired by the University who were trained by IBM. Of course, in those days, IBM was business Tech's Big Dog and, for the most part, it's only dog.

Fast forward about ten years. In 1980, my third child was born (a girl, YEAH!). I was trying to manage 3 children under 5 yrs old and complete a dissertation in philosophy at the same time. I was tough and determined, with an extraordinarily helpful husband, but even I had an insurmountable problem. We were living on academic salaries, and there was no childcare in the budget. There was no way I could carve enough time out of the day to type a dissertation on our electric typewriter, with carbon copies, no less. (There were no affordable copiers available to the public yet, and academia wouldn't accept photocopies anyway).

I tried valiantly to do it, though. With a tiny bottle of Liquid Paper correction fluid ready to hand, I spent precious chunks of time carefully painting over every deletion and typo on both the original AND the carbon copy, trying desperately not to smudge the carbon copy or drip onto the carbon paper. Of course, this painting had to be done immediately after each error was made, without removing the paper from the typewriter or misaligning it. The painting was then followed by a period of huffing and puffing to dry out the fluid on both pages. If the fluid was even a little bit wet, the newly typed correction became an unreadable smear. Oh, I tried so hard and shed many tears, inventing new swear word combinations daily, but the writing was on the wall. I couldn't ignore it. I'd never finish my Ph.D. this way. Even if I somehow managed to finish typing it, the end result was destined to be an unacceptable mess, rife with smeared carbon copies and sprinkled with the indelible presence of baby spit and worm guts from little boys' fingertips. The finish line was in sight, but I'd never be able to get there. I felt so low.

And then a miracle happened. I found a way to type my dissertation without needing carbon copies or correction fluid. Deletions and rewrites were quick and easy, and the end result looked as though the changes had never happened. I submitted the most beautiful looking, typed dissertation ever recorded in that Philosophy department, and I was able to provide any number of perfectly typed copies. I got that damned Ph.D., too!

Tech, or more precisely, SCRIPT, was my miracle (what is Script?[3]). Although I didn't know it then, Script gave me the foot that I could later put in the door to get inside Tech. Here's the story of how this miracle came to me.

It was serendipity that in 1980 the Philosophy Department (where my husband was a professor) was stuffed into the rear of a depressing old armory building, which also housed the university's Data Center. What a strange brew that was. The jungle drums of the Data Center's mainframe sucking in and spitting out cards, combined with its clattering cardpunch machines, provided the mind-numbing background for the adjacent philosophy classrooms, where students were trying to wrap their heads around wildly abstract courses in metaphysics, epistemology, and logic.

In hindsight, I believe this subliminal throbbing of the Data Center drove my logician husband to teach himself Lisp as an experiment for his Artificial Intelligence/Philosophy of Mind course. (Back then AI was firmly in the bailiwick of philosophy[4] departments, although my husband's course was bleeding edge for those times.) He programmed a few Lisp assignments, which a few of his students completed in the Data Center. But they didn't have to punch cards to complete their assignments. Instead, they were given access to one of the few, brand new CRT terminals that were hooked to the mainframe, and created their Lisp programs by typing into the green screen. A few months after succeeding with this initial foray into programming, my husband said to me, "Honey, I want to leave academia. Computer programming is the next big thing, and there's big money in it. We'll be able to pay our bills and have money left over! And guess what? It's FUN!"

And so it was that, in an attempt to lift us out of poverty (except for food, we shopped exclusively in the local Salvation Army store), we agreed that he should get a masters degree in engineering—-specifically, Information Engineering. (There were no Computer Science departments yet at the major universities.) After that, he would switch careers and follow the money. So it was that, by mutual agreement and after teaching all day, off he went in the evenings to learn about data structures, computer architecture, and programming the mainframe. For us, the era of Geekdom began, and looking back, I would say we were then living in 1 AG (Anno Gekum).

The rest of that period started happening pretty fast. After a year or so, while I was struggling with my dissertation problems, he was doing his homework at home using a CRT terminal that plugged into a cat (lol). No worries. This CAT[5] was an acoustic external modem that connected to a home phone and transferred 300 bits per second over the phoneline to the mainframe at the University. No, that is not a typo. Data was sent at the rate of 300 bps (FYI a bit is one-eighth of a byte). "Acoustic" meant that when the children ran in from outside screaming that they found a giant toad, the modem picked up the sound and, well, garbage in, garbage out.

So, here's the kicker. As a distraction from my dissertation misery, I soon picked up some of this Tech, just for fun. In the darkness of late summer nights, while my husband at class, I had fun sitting at the CRT green screen (black screen with green text) with a suckling babe in my arms. With a few fingers on each hand free, I played Star Trek[6] on the university's mainframe. I won more often than not, although, truth be told, some nights I ended up with the Star Truck. Win some, lose some. But one day, when he came home from teaching and saw me bent over the typewriter in "dissertation tears," my husband introduced me to Script on the mainframe. Because of Star Trek, I was already up to speed on the Tech of a mainframe green screen interface, and the structure of Script was so easy to learn. It was a very logical, tag control language for word processing. (Oh Snap! Although I didn't know it at the time, a vision of my Tech future had just appeared to me---can we say HTML?)

In a few days, I was able to burn those carbon papers and step into the new world of Tech. Yes, it was just a black screen with green lines of text, corralled by fixed line numbers and managed only with repetitive, clumsy key combinations. But it freed me. I could close the typewriter into its case and put it under a table. Thereafter, it would only gather dust. No more tears from me. Before long, the time came to print my dissertation on the mainframe, which sent the printout of my work to its daisy wheel[7] printer terminal. Each time I needed to print, Daisy gave me perfect output. To this day, my original dissertation exists on reel to reel tapes in my storage cabinet. Data in dinosaur format.

2.2 *My second step*

In 1983, when I began my academic career as a newly minted Ph.D., our department admin, Dolores, was unpacking the department's new "personal" computer. Like 99% of folk working in academia then, Dolores had no clue what she was expected to do with this item. I walked in to her office to introduce myself, and we were soon unpacking the little beast together. It was a Kaypro II[8] running CP/M. It took both of us to lift that heavy sucker out of the box and onto the nearby file cabinet.

Now what? The department members were expected to use the new computer to help them with their writing. Dolores herself wanted nothing to do with the thing. She had her coveted IBM Selectric for admin duties and desired none other. I, on the other hand, viewed any and all Tech as my miracle BFF. Gazing at the beast, I heard the Gloria of angel choirs. I actually knew words like "operating system," having dealt up close and personal with one during my Script adventure (TSO on top of MVS[9]). Even better, I had actually done word processing. I knew what it was and couldn't wait to tame that little beast!

With no other volunteers in sight, I happily undertook my first Tech job as the department's "IT guy." It was a non-paying gig that no one wanted, but I took that little Kaypro and enthusiastically bent it to my will. I poured through its manuals and mastered its clunky little OS. I slid those floppies in and out of its slot and installed WordStar[10], an exciting new type of word processor (an early WYSIWYG). Last but not least, I made the beast speak nicely to an Epson dot matrix printer.

Keep in mind, dear reader, that this Tech setup was true esoterica in those days. It was most definitely not a plug and play situation. Rather, it was I, the IT guy, wrapped up in cables that were attached to hardware that was sticking ass-up on a desk. Passers-by saw a young woman partially hidden in a bunch of wires and gingerly shining a flashlight into metal guts. They likely heard some muttering too, such as "OK, now what's a printer configuration switch and where are they located, and which ones get toggled up or down under what conditions?" After a few hours working on that setup, I was frazzled.

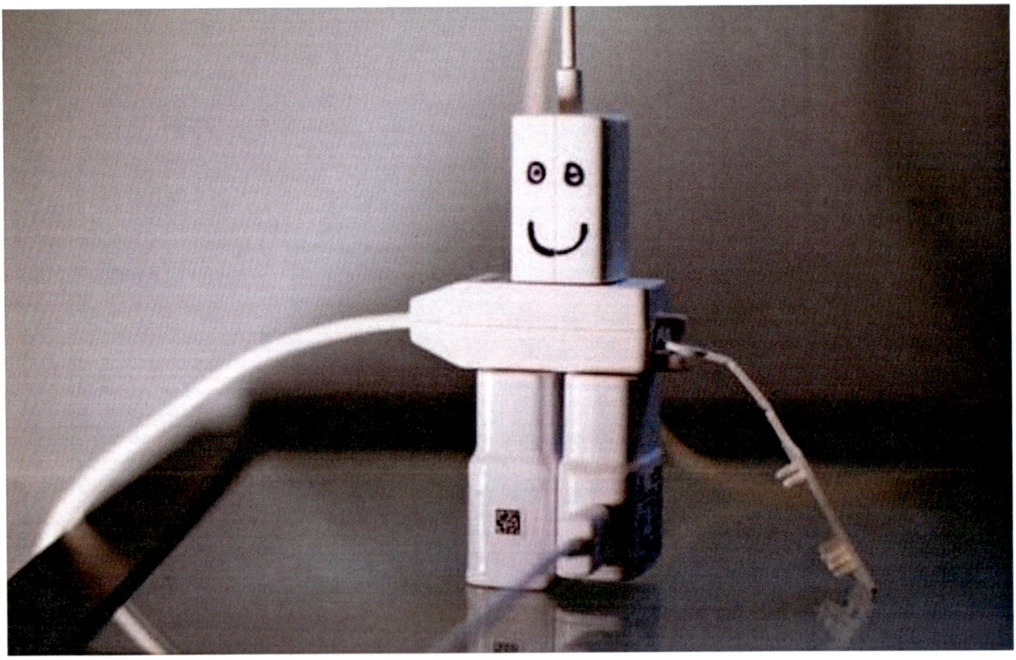

Using such a (now ancient) Tech setup, I "wordstarred" and printed all my academic papers and sent them off for publication. At first, my colleagues ignored the little beast that was clattering away in the extra room down the hall, with its ridiculously tiny screen. But they eventually came to see how much faster I could get a publishable paper written and printed than they could. Keep in mind that, in academia, publication is the way to success. Within six months they each asked me to train them on WordStar. Ever the teacher, I provided them with written instructions for wrangling that awful application as well as what to do if the little beast barfed during a WordStar session. As I see it, this was my first foray into technical writing—a skill that came to prominence in the Cleanup Crew.

Dolores and I became good friends, but she never warmed to use the Kaypro, despite my strong encouragement. As it happened, the next year I turned down the department's tenure-track offer because we decided to move to a rural area in another state. Although I planned to find another academic position there, I had to admit that my adventures as the "IT guy" were a lot of fun and really quite interesting. They stuck with me, but I didn't see them as much more than a good party story then.

As it turned out, it took us a year after that to make our move, and during that time I had more fun writing a short manual for an application created by a programmer at my husband's new job. This time I was paid for that little bit of technical writing—and paid handsomely. Meanwhile, my husband was making money hand over fist writing compilers (one of Tech's Black Arts). Our future was unlimited, bills were always paid, fancy family vacations were taken, and life was good. So far, I had unwittingly accumulated a suite of Tech skills perfectly suited to a Cleanup Crew. I had experience using a mainframe computer OS and the latest desktop OS. I could use the latest and greatest desktop word processor, as well as the old mainframe Script. I could write clear and concise instructions for using a software application. So what? I had no interest in going through that Tech door. I was a successful Professor. I was pursuing a promising academic career.

2.3 My final step

It was 1987. We had finally moved. Our mini-farm in the rural south was bliss. Peaceful, rolling green hills, the scent of pine forests, neighbors too far away to see, kids running about in clean air, fishing in the ponds, building pens and a barn for chickens and turkeys, endless streams of dogs and cats, and my beloved horses. We regularly saw soaring eagles and ospreys hunting over the lake as we crossed it to go into the "city."

Our little country town was only a 45-minute drive to a hotly developing new Tech area, nestled in the farmland of the state's major city. While my husband had landed a sweet job there, my job hunting resulted in an unpleasant shock. Academia was snubbing me—me, God's gift to the discipline! What happened? Well, the market for philosophy Ph.D.'s was ice cold. University students preferred more "relevant" and "practical" courses, and philosophy departments were losing enrollment. I had been out of the game for a year, so no one remembered me. Furthermore, the options in the rural areas were few.

Given the situation, the couple of interviews I had always focused on why I left my previous position. The members of these departments were suspicious of someone who had turned down a plum tenure-track position at a major university—a position they would have jumped at. Maybe they thought I was not committed, or didn't deserve another job, or else just not a suitable fit. Still, I was my determined self and tried for 2 years to get an academic job. It never happened though—at least not on the terms that I considered worthy of me.

During that job search I kept busy running the 4-person office of our local, countrified real estate. They had a fairly new personal computer running MS-DOS, but only the boss knew how to do anything with this PC. I took over that PC, spreading my wings with graphics for the real estate brochures and creating all kinds of little office-based data files. I was their Star, for about $6/hour. There were advantages, though. It was sort of interesting computer work, as I ran my own show. The boss was able to go back to selling real estate, and I was able to be home with the kids during non-school hours. But after a year, I started wondering, "All those years of struggle to get the Ph.D., and Office Manager in Bumfuck is going to be my future?"

Whatever. Life happens. In 1989 we experienced a family crisis, and I needed to get a job. From my point of view, the situation was desperate. Not only was my husband suddenly very sick, but we had a catastrophically ill child and were now in grave danger of losing our wonderful medical benefits. I *had* to get a job that offered suitable benefits, should it turn out that we needed them.

OK, so no time to fool around. What good paying job with benefits could I get RIGHT NOW? In those days, jobs were advertised only in newspaper classifieds. I saw lots of jobs for Technical Writers. I knew what that was. I had done a bit of it. I had Tech skills. So, Tech writing became my primary job target (but I looked for office work too). I crafted a suitable tech writer resume and brought a copy of the short software manual I had written 5 years ago as evidence of my experience. I had many, many interviews. They all ended in "You have no experience, and your Ph.D. makes you overqualified. Good bye and have a nice day." I took the Ph.D. off the resume, and I kept going to interviews. Years later I realized that, although I was devastated each time I didn't get the job, my interviewee skills improved with each defeat. This was a vital lesson for success---learned the hard way.

After 6 months of anxious trying, I found a great entry-level tech writing job with a contracting company that worked exclusively with IBM. My interviewer actually knew what Script was, and it impressed him when I said that I had used it. He was impressed with my little manual as well. He didn't like that I had no recent experience, but he was extremely anxious to fill all 5 of the positions for his new project at IBM. He took a chance on me because I interviewed well, because I had a sample of my tech writing to show him, and because I claimed to know Script. He figured I might have a clue about IBM mainframe Tech, which he needed for his project. Most of all, though, he hired me because he needed to fill his positions quickly, and I was lucky enough to be in the right place at the right time. Looking back, I see that I had made my luck by stubbornly continuing to interview after so many failures. In any case, all I cared about after that interview was that I had finally found a job good enough to assure medical care for my family, if the worst happened.

In terms of my future, my foot was now tentatively in the door of Tech. It would be up to me to walk through that door, where money and opportunity were waiting. For sure, I got lucky finding that job, but more important, I made my luck happen. You can do it, too, dear reader. Never dismiss the Tech skills you have accumulated and the Tech products you have created while on the way to doing something else. With persistence, you may be able to fashion them into stepping stones to a new and better career.

And now, as we say down here, I'm fixin' to tell y'all about my wild ride inside Tech and some of the characters—good, bad, and crazy—that I rode with.

CH 3. Groupie To the Early Geeks

Ahhh, the Golden Age of IBM. Over a 4-year span in the early 90's, I caught the end of IBM's Golden Age. Yes, its Golden Age was fading, but because IBM was so big and so very rich, the end came slowly. During the 90's the company still had money to burn, and I, like thousands of other contractors, took lots of that money to work on months-long projects that were never really required to be successful. Though the projects were relatively unimportant to the company, they offered me the opportunity to work with bright people and learn a wealth of Tech. On IBM's dime, I developed deeper and broader Tech knowledge and high-level Cleanup Crew skills, which boosted my worth and lead to better jobs. But sadly, I also watched as the world's first, amazing geeks failed to keep up with the pace of Tech and were allowed to make a number of wrong turns. Those powerful, elite geeks, who had made so much great Tech (my miracle Script, for example, which was just a minor exercise for them), crashed and burned the IBM car. The company I see today is a grotesque knock off of the Big Blue I knew. The usurper now flashes the great one's name to gain whatever fleeting attention may be left.

Such is the way of the Tech business. When you witness the fall of a great company from the inside, you learn that it can happen also to the Apples and Amazons, the Googles and social medias. At the same time, you learn that other giants will arise because Tech never stops. Tech will never stop unless humans stop, and opportunities to jump in will always be there.

But the early, great geeks from IBM set the standard for later generations of their kind. They embodied the character and mindset that Tech could later hold up as the best of the best. So, who were these guys, and what were they like?

3.1 They don't like to talk

The IBM that I worked for as a consultant committed itself to excellence in process and unequivocal adherence to high standards for deliverables. There were no cowboys in this company. There were specific ways of doing things, and that's how you did them. Why so picky? IBM sold "big iron" that had to handle huge workloads of critical path data processing. Banks, car companies, airlines, governments, stock markets, big pharma—-they all depended on IBM's mainframes. Big iron was the company's bread and butter. Customers happily and repeatedly paid millions of dollars for their mainframes. These behemoth machines and the software that ran on them had to be secure, reliable, maintainable, and updatable. Such requirements were not, and could never be, optional if the company were to continue to succeed and grow.

The IBM way ensured security, reliability, and repeatability in the manufacture and development, and deployment of IBM big iron and its software. The IBM way worked for a long time and was the inspiration for other companies. The project cycle, the management process, the standards for a project's deliverables (whether hardware, code, or documentation): these were all clearly defined and available in IBM's internal policy books. So it was written, so it was done, from IBM's critical mainframe projects down to their less significant or irrelevant projects.

Irrelevant projects? If your project was not in the critical path of what big iron needed, it was irrelevant. The company was enthusiastic about its irrelevant projects but also OK if they never saw the light of day. During its Golden Age, IBM took pride in funding all kinds of irrelevant projects. If such projects kept the company's bloated personnel rolls busy, allowed them to try something new and innovative, or contributed to the company's PR narrative in some way, these irrelevancies were worth the money. And there was so, so much money that keeping careful track of how it was spent inside these irrelevant projects did not have high priority.

When large or small, often irrelevant projects were underway, it was common for IBM to hire contractors to do the work. This allowed management to reward the IBMers with the role of boss. These IBM bosses of contractors had a special name: Vendor Coordinator (VC). In the Golden Age, there were many VCs who were looking for contractors to staff their projects. With such a high demand for contractors, it's no surprise that the body shops (technical recruiting companies) latched on to IBM like ticks on an elephant. My first technical writing contract job was part of this hiring frenzy. IBM had decided that all of its documentation for the big iron needed to be put into a word processing application, rather than sticking with the cumbersome type-setting process for printing books. Word processing was the new Tech, and IBM was going to use it. While the big iron segment of the company didn't desperately need this project, it could potentially benefit from it because documentation would be faster and cheaper to update and print. So, my first IBM technical writing project was more relevant than irrelevant and had great funding.

I had used WordPerfect and WordStar for word processing on early desktop computers, but that isn't what IBM was going to use. To ensure that their new process for book making would support the highest documentation standards, broad functionality, and, of course, work on the mainframe, IBM decided to create its own text processing application. So it was that IBM BookMaster came to be. In my new position as a technical writing contractor at IBM, my job was to become proficient in BookMaster (fondly referred to as Bookie), and then transform my assigned big iron books into Bookie format. This meant that each hard copy book was to be manually typed into Bookie files with the appropriate Bookie formatting codes applied. At the end of this process, the Bookie book would be printed and edited to ensure that each page looked exactly, precisely, like its hard copy original. No mismatch of even a period, a comma, a numbered list, an extra space, a chapter title's font and style, or the position of a page number would be tolerated. If we contractors could not live up to this standard and perform our tasks on schedule, the word would come down from the IBM VC to our boss, "You're fired." There were too many contract companies banging on IBM's door for the VCs to tolerate the least bit of incompetence.

On my first day of work, as the recruiting boss explained the project details to me and my 4 new colleagues, high anxiety vibrated through every one of us. The young woman sitting next to me was literally shaking. She was a recent graduate with a Master's degree in English, and this was her first Tech writing job too. Could we do it? We would be given 8 weeks to become proficient in BookMaster, and our training would start the very next day at 9am sharp. Our training room at the recruiting shop would be visited occasionally by our VC for monitoring purposes, so no absences would be permitted and business attire was required.

I was a bit nervous, but still so happy to have this new job that I slept well that night. We began training the next morning in a stuffy, crowded, cave-like room in the recruiting offices. After hours of tedious preliminaries, mostly about IBM's security rules, we were finally introduced to our IBM 3270 green screen terminals. We were shown how to log in to the MVS operating system, which was confusing to my colleagues but was a no-brainer for me, based on my old Star Trek and dissertation writing days. Next, with some fanfare from our trainer, we were shown how to invoke BookMaster and create a new file for text entry. We were then given the manual that described the zillions of tags to use for formatting the text we would be entering. It was nicknamed the "Bookie bible," and we would have to master every word of it. My fellow contractors sucked in their breath. They were scared. This manual looks like a programming book! I quickly thumbed through the Bookie bible and immediately almost fell out of my chair. All of the formatting tags were totally familiar. Bookie was just Script with a new name!

The rest is history. I suffered through the boring training, and then we started work in the hallowed 2nd floor offices of building 5 in the nearby IBM complex. Gee! Wow! Look—there's a huge cafeteria and free coffee and the ladies room feels like a spa! Everyone is so friendly!

I quickly became a Bookie star while luxuriating in building 5's amenities. I performed the most complex Bookie tasks. Really, I could do this work in my sleep. Furthermore, I could figure out why Bookie occasionally barfed, and how to fix it. Also, I could clearly communicate "the why and the how" to others who needed help. (These were all skills I had accumulated during my Script and Kaypro days, but had put aside as not in my critical career path.) Soon enough my work was being noticed by the IBM VCs, and their interest in my skills was filtering down to my recruiting boss. He recognized the potential. If I could work at a higher level, he could sell me to IBM for a lot more money. A good raise for me and a whopping 30-50% of a much higher hourly rate to him. Win-win.

After less than a year as a new contractor, I made a big jump in salary and job status. I was being sold as a Senior Technical Writer. Only a short time had elapsed, but I had changed. In contrast to my academic way of writing, I had learned the IBM way of technical writing, and it was magnificent. It demanded clear, concise, organized writing that drove quickly to the point and provided all the useful information the targeted readers needed but not a word that they didn't. No dithering, no extra paragraphs of possibly related but purely background knowledge, no throwing in the occasional subtlety or irony to break the somber academic mood. Short sentences, straightforward paragraphs, simple, direct, and complete instructions—those elements were my writing style now. Even more, the appearance of a page was expected to highlight the meaning and purpose of the page. Formatting and style were not viewed as independent of content. Elements such as numbered headings, heading levels, sections, lists, tables, graphics, and blocks, which we could control so thoroughly with Bookie, were our tools. We used them to bring content to the reader in the correct flow, to show critical relationships among statements, and to highlight instructions with the correct level of importance.

Our work was constantly scrutinized, too. I confess that I initially struggled with the IBM editing process. After all, I was a well-published academic. I knew how to write and assumed I didn't need help from an editor. But at IBM, no book was published without an editor's sign-off. IBM's cadre of editors had advanced degrees in English, composition, and information science. They were familiar with the content of all of the books in the IBM mainframe library. In addition, they understood the target audiences for these books and the role the mainframe systems played in the company's portfolio. When they edited your book draft, they expected it to be "ready for edit." That meant no spelling errors, typos, or formatting errors. They were not proofreaders. They edited to the IBM standards of clear, organized, reader-targeted, and concise text. If the way you wrote could be or should be better, they marked it, told you why, and gave an example of a better way. When you received your edits, you made the changes, period. No delaying, no argument, you just made the changes. IBM backed their editors 100%. If you didn't heed an editor's markup or disagreed with it, your contract was not renewed. IBM believed that simply having standards is not sufficient for quality results. Standards must be enforced. They were right.

What I learned at IBM was not just good technical writing, but great technical writing. IBM documentation was known to be the finest available and highly prized by its customers. They relied on it and guarded it from borrowing-minded colleagues, typically dedicating a special bookshelf to their IBM books that was close by their desks. The IBM way of tech writing set a high bar, and I strove mightily and proudly to meet it.

As a newly ordained Senior Technical Writer, I was told that several VCs at IBM wanted me for projects—irrelevant, maybe, but still high-level projects. These projects required a rare skill, which they believed I might have and were willing to pay the bigger bucks for. Specifically, I would need to work directly with one or more IBM developers. I would be writing brand new documents describing the Tech these developers were making. I would be creating introductory manuals, installation manuals, and user manuals for the new Tech. To accomplish all this, I would have to talk to the developers. This is where that special, rare skill would be needed because, well, developers don't like to talk.

They never put it exactly that way, but I soon understood that's what they meant.

3.2 Elegant code

Between 1989 and 1993, all of my contracts were with IBM. About half of those contracts were with fairly high-level development groups. Among them were several with the Watson Research Center, one with the Networking Center in Rome (Bella Roma, as we called it), and a couple of what IBM called "skunk works" projects in various New York state locations. Sometimes I was the only tech writer on the project, often by request. But most of the time there were 3-5 of us working on an entire library of documentation deliverables. IBM paid for our travel to meet with developers over the course of days, weeks, or months, as the project required. The travel budget was not lavish, but it certainly was above average. We writers always enjoyed our trips, but I was typically the only one who enjoyed working with the developers that we met.

My colleagues often complained that they could not get any information out of the developers they were supposed to work with. A tech writer's job is to describe a technical system and provide appropriate instructions for using it. But first, the writer must understand the system. Only the developers who make the system can provide the requisite level of understanding that the writer needs. Unfortunately, the initial meetings with said developers usually started with the writer asking a few introductory questions about the system, and the developers responding that the answers are in a requirements document, or a specification document, or some other document that they have already provided to their management. What wasn't said but was clearly implied was that they were not about to repeat this information. For the most part, writers were stopped cold at this point, and developers started to glance at their watches. I came to recognize this watch-glancing behavior as their way of saying "You are now wasting my time." While they were happy to talk sports with each other for long intervals during the workday, they sure as hell were not going to sit in meetings with people who were clueless about what they were working on and who seemed to expect a tutorial from them. Wasting a developer's time was about the worst thing anyone on the Cleanup Crew could do. They knew it, and if we didn't know it, we learned pretty fast.

These initial, intimidating meetings set the tone, and writers subsequently found themselves in a constant struggle to get time with their developers and pull the needed information from them. Work progressed slowly and early drafts of the documentation were filled with holes, notated with the writer's comment that the requested information had not yet been provided. IBM VCs did not like to see moth-eaten drafts and began their own cajoling and whining to the development team to please give the tech writers what they needed. All of this was ignored by developers, and they eventually would delegate the task of talking to the writers to the least important among them. This unlucky person would become the intermediary, giving information he thought was correct to the writers and then asking them to rework it after the developers read the drafts and told him it was wrong. No one but the developers liked this ridiculous dance, but no one dared to complain about it. VCs only shrugged and said, "Well, you know, they don't like to talk."

My approach to a new project was quite different from most of my colleagues. I wasn't afraid of learning Tech on my own, at least at a 50,000-foot level, and I found even the nerdiest of it interesting. I also tried to put myself in a developer's place to understand how he viewed his world. I created a developer profile that grew in detail and refinement with each project experience. I began with the idea that before I met with a developer, I would read every document available from management for the project, both marketing and technical. As I read these documents, I would hit the library and research any terms I didn't know (no Google in those days). Based on what I learned, I would put together a picture of the system I would be documenting. Immersing myself in the project documents that I had, I would highlight the sections I had questions about and write notes to myself. Even after about a week of this effort, I would still be very ignorant of the system. But at least I knew what was knowable at my level and light years ahead of my colleagues, who walked into their meetings with only the VC's 15-minute overview of the project.

After my preparation by immersion, *my* initial meeting with developers was a horse of a different color. First, they saw me sit down and plop a pile of their documents on the conference table. Then they noticed that these docs were peppered with yellow highlights and sticky notes. They would sit back and give me a look. "Has she actually read them?" would be all over their faces. After that, I would start asking my questions. For example, (substituting *x*, *y*, and *b* for the names of real features) I would say:

> *As I understand it, the system does* ***x*** *and* ***y****, as described in the requirements document. But the functional specification document seems to describe an additional feature* ***b*** *of the system that isn't listed in the requirements. Can you tell me if there is an additional feature* ***b*** *of the system, and if so, how it's different from features* ***x*** *and* ***y****? Also, if* ***b*** *is different, why wasn't it mentioned in the requirements?*

Now the room would ripple with interest. There was no watch-glancing. Here was someone who had a clue and had clearly tried to understand their work. They loved it! They couldn't get enough of it! Most important, now they wanted to talk. The meeting would go fast. I would have so many questions. Each of their answers would boost me to the next level of understanding of the system, and because of my preparation, I could usually follow up with a more advanced question. I loved the learning, and it showed. They loved seeing me learn and liked to keep throwing information at me to see if I "got it." After a few scheduled meetings like this, I would be invited into their inner sanctum. I could walk to a developer's office and ask if he had time for a quick question. Rarely was I turned down. On many occasions I would walk into an office where several developers were discussing some highly complex aspect of the project, talking about the parts they were working on and the difficulties they were having. They let me listen, and even though I didn't understand their issues, in those moments, I could see into their souls.

The best of them had minds that would race down logical pathways, creating decision trees of possibilities for the code, sorting and pruning solutions according to completeness, feasibility, and, last but not least, elegance. They had severe requirements and constraints. In addition to performing the tasks the customer needed, the code had to work reliably and securely. It had to be upgradable yet work on both old and new computer hardware. Finally, the code had to be ready to deploy on schedule.

It was a high bar, but par for the course for developers. Those of high caliber took all this in stride in sort of a ho-hum way. Yes, of course they had to write code that worked. But that was barely a challenge. In fact, in every project there were always any number of code implementations that would satisfy the requirements. What excited, inspired, motivated, and fulfilled the good ones was the possibility that this time, for this project, in a big or small way, they might be able to invent *elegant code* that satisfied all of the requirements and constraints.

Elegant code does things in a way that is unexpectedly simple yet powerful. It makes other processes easier to implement or eliminates the need for them. Sometimes the elegance comes from a completely new way of approaching a coding problem. Sometimes it comes from taking a well-known approach but using it in a new way. To read it, understand it, and watch it work is a thing of beauty. The experience is similar to viewing a work of art. For the developer who creates the elegant code, the effort is not in the realm of what technicians or scientists do. The developer does not call upon those kinds of skills to create elegant code because it is, fundamentally, a work of art. It isn't achieved through methodical processes. Elegant code is about imaginings, foolish mixtures of ideas, odd intuitions, and those moments of insight that come in the night. It is a dance between the developer and the elements of a programming language, as the artist dances with color and light. Elegant code is the work of the engineer as artist.

As we often see with artists, there is a sadness bound up in the creation of their works. It's no different with elegant code. The developer knows that the code will only ever be appreciated and understood by a very few people. But even worse is the fact that the target audience for the elegant code—the customer/user of the software—not only will never see this beautiful code but also doesn't care if it's there or not. Customers and users care only that the code works and the software meets their needs. They don't care if the code is everyday hum-drum, a million lines longer than it needs to be, and something only a mother could love. They will never know or care if they are using a work of art. Sadly, the developer's drive for elegant code is a lonely road.

Is it any wonder, then, that they don't like to talk to people who can't understand or appreciate elegant code?

3.3 Extreme perspicacity

You can't bullshit the code. Code either works or it doesn't. If it doesn't, you either fix it or throw it away. There is no in between. Broken code has no value. It is not like an interesting idea that didn't quite fulfill its promise but is still worth keeping around. You can't try to talk code into having value. You can't provide a story, whether true or false, according to which developers will be persuaded that broken code is worth something. You can't bullshit the code.

This strict nature of code forces developers to exist in a world where only what is clear and demonstrable is reliable. In the time of the early geeks, developers held code reviews. In these meetings, a developer's code was reviewed by peers and management, line by line. What the code was supposed to do and how it was supposed to do it were judged. If, upon inspection, you could not explain clearly how the code worked or how it would do what was expected, or why it was a good solution, there was no mincing of words from the reviewers. Obscurity or redundancy of code was not tolerated. Mistakes were ridiculed, "Why didn't you see that problem?"

These review meetings were harsh, and yet, in a strange way, they were refreshing for the good developers. They emerged knowing whether their code had problems and exactly where the problems were. They were usually given some ideas for solutions. After a code review, developers knew the extent to which their mistakes were stupid and the extent to which they were plausible.

It was personal, but at the same time, developers knew that each of their colleagues would have to go through the same bath of fire with their own code. During a code review, the targeted developer experienced the pure rationality that comes with demanding clear, clean code and settling for nothing less. No bullshit could get by that group. Any weaseling or persuading was immediately called out. It was a relief not to have to try to make excuses for your failures. It was an honor to participate. The atmosphere in a code review was fresh, rarified air.

If you could stand up to that kind of environment, not only did you come to like it, you also endeavored to recreate it in all of your interactions. What kind of intellectual background did these developers come from? Universities didn't yet have computer science departments. The ones I worked with were trained as mathematicians, or chemists, or physicists, or electrical engineers. They were steeped in precision and rule-based thought processes. There is no wiggle room in a mathematical proof or the equations governing the operation of a machine. Given this way of looking at the world, it was no surprise that their workplace behavior was grounded in exceptional (some might say over the top) intellectual rigor.

Unfortunately, the world outside of the developers' enclave was not prepared for their perspicacious behavior. Any idea, claim, assertion, or opinion brought forward by a member of the Cleanup Crew during a project meeting was fair game for the developers who were present. They might challenge and pursue any word or phrase in a draft document, vigorously arguing until the meaning of it and its implications were laid bare. Every suggestion or comment was subject to evaluation for clarity and feasibility. Only then was it worthy of acceptance or rejection.

For example, if a tech writer's draft contained a paragraph or even a word or phrase that most would say was understandable although perhaps had some minor ambiguity, the acceptance of it could become a heated discussion that went on way too long and seemed unnecessarily critical in tone. In these situations, the best of the developers honestly intended no harm or disrespect, but the lessor of them sometimes engaged in the behavior because they enjoyed intimidating others with their skill of argumentation. In any case, their extreme shrewdness was often offensive and unpleasant to others on the project team. It also led to mockery and name calling by members of the Cleanup Crew when the developers were out of earshot, "Those weirdo nerds, no wonder no one likes them! I wouldn't want to be caught dead in public with one." Thus began the unflattering mystique of the geek.

I had to agree that they were obnoxious at times, but I understood that a zebra cannot change its stripes. To succeed in their careers, to create elegant code, good developers have to embrace all and only that which is clear and demonstrable, and challenge everything else. By necessity, this is their way of being in the world. Their souls are a mix of engineer, artist, and geek, and they make amazing Tech. After realizing all this, I thought to myself, "Well, that's really not so bad, and who among us has the perfect personality? Let him cast the first stone." I also had to admit that it was great to be around people who don't tolerate bullshit.

3.4 High-speed innovation machine

Technology marches through our human history without interruption, inexorably advancing our ability to control our environment. When you stop to think about it, you realize something remarkable about the part of technology we call Information Technology (what I'm calling "Tech"), namely, that it advances so much faster than other technology. Why?

Innovation is the force that drives technological advancement. In my time inside, I saw Tech innovate at an extraordinary pace when compared with other technology. It occurred to me that, unlike other technologies, Tech flies free of the physical world because the "stuff" it works with is code. And code is nothing but thought instantiated in symbols. Code is not bound by material properties. It is not subject to the laws of chemistry, biology, and physics. Code is like ideas, infinitely malleable, or close to it. Contrast this with the "stuff" that other technologies work with—-for example, medical, industrial, transportation, and civil technologies. Their advancement can occur only within the formidable constraints of physical material: DNA, calcium channels, nerve growth, O rings, temperature, pressure, wavelengths, kinetic energy, melting points, chemical bonds, tensile strengths, etc. Options for new constructs are much fewer with physical material. Innovation is harder to come by, no matter how great the thinker. Getting outside of the box can be impossible for a project. Progress is inevitably slowed by the time it takes to uncover, understand, and manage the incredible numbers of constraining variables involved in making a change to physical material.

Making radical changes to code, however, requires none of that laborious procedure. Relative to other technologies, the Tech mind can run at high speed when pondering radical changes because code has no physical limits. Further, once radical changes are conceived, they are relatively fast to make and easily tested and propagated. You might think that Tech innovation is constrained by hardware. Yet in my time, I saw software innovation drive hardware innovation. Innovative code stimulated the invention of hardware that could support and enhance it. Hardware played a small role, if any, in holding up radical code changes. So, the possibility of frequent, radical code changes always plays a strong role in Tech's more rapid innovation. But from what I experienced, there is another ingredient required for Tech innovation, and we've already met it: fiercely rational minds, with no tolerance for bullshit and an artist's desire to create elegant code.

We know that elegant code always brings something new to the table. Within every instance of elegant code lies the germ of Tech innovation. This germ may be developed immediately into some amazing, life changing Tech. Or, it may lie dormant for a time and be energized under new circumstances. The dramatic rise of Tech that I witnessed happened because elegant code just kept coming, faster and faster. And innovation came with it. Every software project I worked on that had at least one excellent developer, in the mode of the early, great geeks, had a good chance of producing elegant code.

In my early days at IBM, for example, I saw it happen over and over again. In 1990 I was the tech writer for an IBM e-book project for the PC. This was when the idea of e-books was still just a speck in the eye of Tech. IBM, surprise surprise, had already done e-books for the mainframe. The mainframe developers had taken the BookMaster hardcopy libraries for the mainframe and created a conversion program that displayed the BookMaster print output online. The converted Bookie files were sent to the mainframe's 3270 terminal screens for display, where they looked just like the printed version. The benefit to customers was huge. Instead of printing out a hardcopy book using BookMaster files, you could run the new conversion program and view the e-book online. The IBM developers named that conversion program

BookManager. Of course, the only place you could view the e-book was on a 3270 green screen display terminal that was connected to an IBM mainframe.

IBM created BookManager to help their Big Iron customers and IBM's support staff deal with the extensive libraries needed to operate and maintain the mainframe. Putting these libraries online---with search capabilities across books, no less---made locating Big Iron information easier and faster. But BookManager was never intended to be a standalone product of IBM. It was only a support product for Big Iron. IBM was the first with a great e-book implementation but never cared much about that aspect. Providing support for its Big Iron customers was what mattered.

The original BookManager was a fairly direct conversion from BookMaster because the code that did the conversion only needed to work in one operating system environment, i.e., the mainframe. The project I worked on, however, was to make BookManager work on a PC running the IBM OS/2 operating system. Someone had the idea that it would be pretty cool to have e-books on these new IBM PCs. Desktop PCs interconnected on office networks were fast becoming the favorite computer setup of business users. Working with applications that ran on your desktop computer was easier than working on the green screen 3270 terminal connected to a mainframe. So, the IBM developers on my new project were desktop guys who also knew all about mainframes. They figured that IBM's mainframe business customers who had to work with mainframes would prefer to read the mainframe e-books on their desktop computers instead of their 3270 green screens. In those days, it was common for business users to have both a desktop computer and a 3270 green screen terminal on their desks, and they used both every day.

When I came on to the project, the developers had just finished creating the first version of the new software: BookManager-PC. They demonstrated it for me, showing a book from one of the mainframe libraries displayed on an IBM PC. Their new BookManager had two parts: a conversion program and a viewer for displaying a converted book on the PC. I thought it was pretty cool and important new Tech. I could easily imagine a future where e-books on personal computers replaced hardcopy. But from IBM's perspective, the project was just a skunk works idea, mostly irrelevant to Big Iron. Nevertheless, in the Golden Age, IBM was happy to fund it.

My job was to write the user's guide for the new BookManager-PC, but with a twist. The book I wrote would be provided as a BookManager e-book on the PC. My writing process was a bit confusing. When I was ready to write the user's guide, I would log into the mainframe and use good old BookMaster to create the book files. When the book was edited and ready to be sent to the printer, I was to shift over to an IBM PC that was running the new BookManager-PC software. Working on the PC, I was to transfer my BookMaster files from the mainframe to the PC, then use the new BookManager-PC software to convert those files to an e-book for the PC. Finally, I was to open and review my new user's guide e-book using the new BookManager-PC viewer. I was to proof this online book and verify that the conversion was displaying correctly, without visual errors. The e-book was expected to look just like a printed version of the book would look. Essentially, I was writing a user's guide for software that I actually had to use to create that user's guide.

From what I could tell, I was the first writer to create an e-book for the PC and learn all about the e-book conversion software itself. As part of my work, I also learned how to code and view hypertext links within e-books, which was bleeding edge stuff for tech writers. Few people in the Tech world had ever heard about such things. The whole project was very cool for those times. Yes, this was a great project for me, and the new BookManager-PC software worked like a charm. I came to understand why. It was made with elegant code.

I spent a lot of time on my own working with the new software, and, as usual, came up with a list of questions for the primary developer on the project. We got on well. During one of our Q&A sessions about the software, several other developers stopped by. They had heard about something my guy was doing with the code, and they asked to see it. It turned out to be a big deal for them. They got excited about it and started talking about using the approach in something they were doing because my guy's way was so much more *elegant.* I remember that they used that word, too. Elegant code was driving the innovation of e-books on the PC, and, as a member of the project's Cleanup Crew, I was a part of it!

What IBM did with BookManager-PC after I moved on, I cannot say. I never heard much more about it, even though the developers on that project believed that IBM could successfully sell it as a standalone product in the PC e-book space. Of course, the software did have one major flaw—it was a PC application, and that isn't what Big Iron is all about. BookManager-PC faded away as Acrobat Reader ascended and took over the space. But that project, with its elegant code and innovation, certainly gave me a leap ahead in skills and knowledge. And I wasn't done yet. I was fortunate to be part of a few more IBM projects marked by elegant code and innovation. Here is one of my favorites.

In 1992 I worked on a high-profile project with developers from IBM's Watson Research Center. There were 5-6 tech writers assigned to this project, and we travelled often to the IBM facility to have face time with the developers. These Watson guys were some of the smartest and most irritating our group of tech writers had yet encountered. Their heads always seemed to bc in code mode, and they *always* knew the answer to everything. But I had no problem with them. They had created extremely powerful computing and visualization software for scientists, and I saw this as a very exciting Tech opportunity.

With this new software, scientists could actually configure the processing operations they wanted to apply to a large data set and then, when the processing was complete, they could visualize the results. One of the system's marketing examples was the configuration and visualization of meteorological models of weather events. The demo given to us by the developers showed a full-color, active tornadic cloud configuration on the screen. The data set behind what we saw contained information collected from real tornado clouds. The software was displaying a visual interpretation of that data, along with a set of numerical properties belonging to the data. The numbers for these properties could be changed on the screen. This configurability allowed users to set up different values for the properties so they could experiment with different models of the original data.

After using the software's graphical UI to change various properties, the users sent the modified data set for processing. When it was finished, the software displayed the new cloud formation with its new property settings. The users could then see how the tornadic clouds changed under the modifications they had made to the original conditions. The visualization was magnificent. We could actually see a tornado in action on the screen within its layers of storm clouds. This kind of software is common now, and scientists use it heavily. For example, we've all seen computer visualizations of DNA, dinosaurs, chemical bonds, the human heart in action, and so on. But with this IBM project, the ability to modify extremely large sets of experimental data on a computer screen with a graphical UI, process the data quickly (for that era), and then watch a visual representation of the updated data on your desktop computer was an amazing innovation for its time. And it required bleeding edge software and hardware that had exceptional horsepower, in a small (for that era) package.

The desktop software ran on the new IBM RISC workstations under the IBM AIX operating system. The RISC architecture was actually designed by IBM in the 1970s, but since it wasn't big iron, not much happened with it until other Tech companies started to make powerful RISC desktop workstations. That woke up IBM enough to fund projects to compete in the RISC-based workstation space. This new software from Watson, for example, was intended to show off the speed and capability of the IBM RISC setup.

The workstation software in this project did the eye-catching visualization tasks, but it was not powerful enough to do the back-end processing of the extremely large data sets. That was the kind of task that larger, interconnected mainframes called "supercomputers" would be expected to do. But for this project, the Watson guys created an innovative, smaller version of a supercomputer that could handle the colossal processing demands. This new beast wasn't a mainframe, but it could kick the same butt as the Big Iron did.

While some of my colleagues on the project spent time complaining about the "snotty developers," I was in heaven. I was the writer for the back-end processing system. Its computing power and speed were tremendous for its time, and even though it didn't take up an entire room like Big Iron, it wasn't exactly small either. They called it a supercomputer; in private, I fondly called it "the refrigerator." I guess it depended on which end you were looking at.

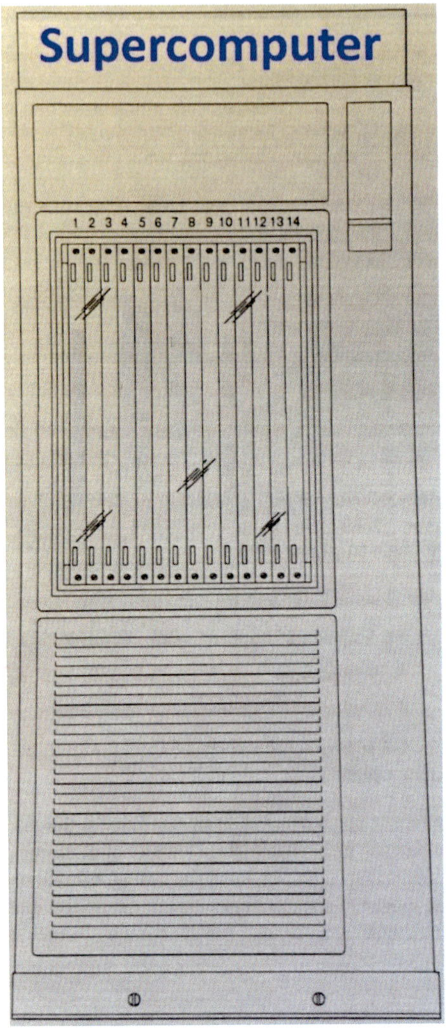
Supercomputer
1 2 3 4 5 6 7 8 9 10 11 12 13 14

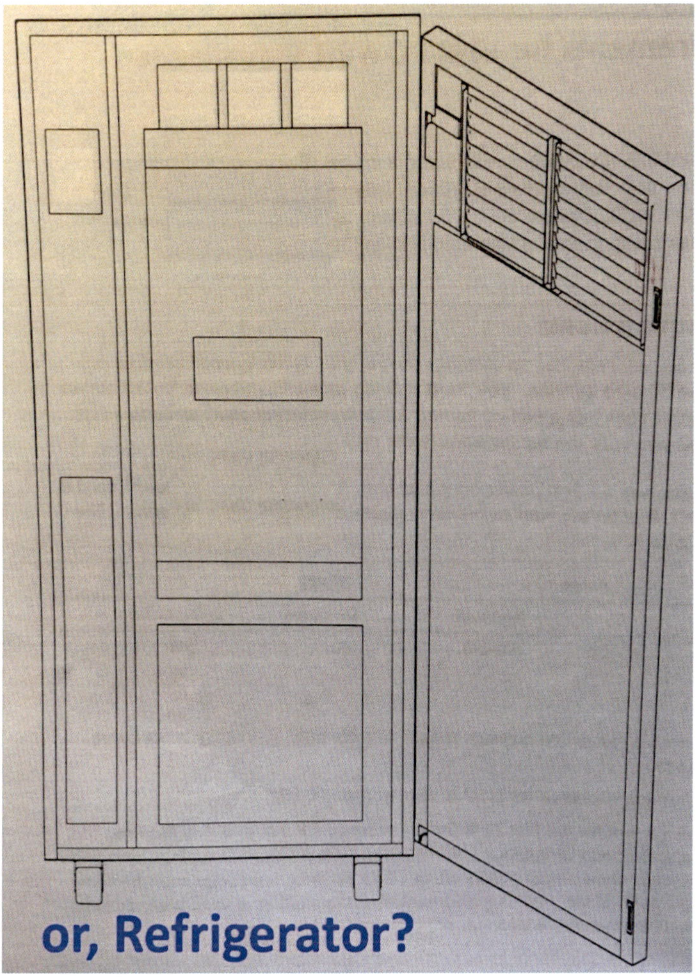

But what was not to love? The entire system was one big hunk of elegant code and advanced hardware. The innovation within it set a standard for visualization software. When the project completed and the system was announced, I know it made a big splash. But after that, I never heard much about it. I do know that IBM gradually gave up competing on the RISC platform. After all, it wasn't Big Iron.

Sometimes, when I reminisce about that Watson project, it makes me think about the many Tech projects happening now and how many more will happen. I think of all the great Tech developers we have and all we will have, pushing out elegant code at high speed. I think of the many Tech companies that thrive on releasing their innovations to the world—like Apple, Google, and Amazon. Without a doubt, innovation flows through Tech, and it keeps accelerating. Fueled by creative developers writing elegant code, Tech is truly a high-speed innovation machine. And, I conclude, that's why it can advance so much faster than other technology.

3.5 A woman among men

I've tried to describe what it was like working with these often brilliant, early geeks in the Golden Age of IBM. For me, it was an amazing experience. But in one specific way, it was also pretty weird. When I was the solo tech writer on a project and met with a developer or a team of developers, I always sensed the presence of a big, fat elephant in the room. Anyone looking in at us would quickly notice that only one woman was sitting at the crowded conference table. At first it seemed so ridiculous that 90% of my project meetings were with men only. Sometimes, it was also uncomfortable. But I grew used to it. There just weren't any women developers around.

The first woman developer I met was in 1995 on a project with the "best software company in the South." This lady was actually a development manager, and her all-male team of developers revered her. She was as smart as they come and great to work with. It was 2005 before I met another woman developer. I was working on a project at a major pharma company. She was the team/project manager and highly-respected by the developers on the project. Then in 2008, I met several woman developers while working on projects at a retail software company. One of them was of the highest caliber, one was a newbie, and the other was average. The development team at that company had a gender-balanced workplace environment, and our project meetings were so much more fun.

During my 25 years in Tech, I was glad to see the number of women developers increase. And I observed women clearly demonstrating that they are just as capable as men of excelling in this career field. Yet, it seems to me that there are still fewer women developers than one would expect. Why?

Socio-cultural restrictions put on women certainly carry blame. They created huge obstacles for women to overcome. It unnerves me to remember that my own mother was born before women had the vote! I don't think those restrictions are the whole story, though. I have an additional thought in that direction, based on my personal experience and my observations of the few women developers I did meet. I offer it here for the reader's consideration.

All the women I met were married and either had children or were planning to have them. However, none of them expected or wanted to end their careers to be stay-at-home moms. They were making the family-career balance work, but it wasn't easy. It took an emotional toll. While on the job, there was always an unavoidable preoccupation with the children, especially if one of them was ill or having some problem. Conversely, while at home, there were always demands of work intruding on family time.

Of course, working fathers experience similar stresses, but working mothers typically experience these stresses with greater physical and emotional pull, particularly when their children are young. For the young ones, Mom is, by nature, the parent they go to for succor and consolation. That's just biology. For the women I met, the solution to managing this high degree of stress was not to interrupt their careers and stay at home with the kids. Instead, they created a pro-family solution that allowed them to keep their careers.

Such alternative solutions were not easy to come by in the early days of Tech. There was no decent maternity leave, let alone paternity leave, no easily accessible, quality childcare, and little tolerance for flexible work hours. These limitations reinforced the view women have always had about career options. This view is based on biology and the social constraints that reinforce it (and men have their own version of it). When a young woman thinks about what she wants to do with her life, she foresees a wide-open period of opportunity until she starts to look past 30. After 30, she says to herself, what about marriage and children? If she has no interest in those options, then her opportunities remain open for career choices. But, if like many (if not most) women, she wants to leave the family option open, then she has to think about its potential conflicts with her career choices.

Looking ahead to marriage and children, her best options are careers that allow for interruption. For example, careers in teaching, nursing, and administration allow flexible scheduling and long stretches at home, or else can be stopped for long periods and then resumed without restriction. Is it any wonder that women are drawn to these careers? They are certainly demanding careers, but also fulfilling, and they allow women to have more of the family life they desire. The careers of doctors, lawyers, politicians, police, CEOs, engineers, and developers have always been more difficult choices for women who want families. These careers require long-term, uninterrupted and on-demand commitments, if you want to succeed and advance. They often require large sacrifices of family time, as well as a spouse who is willing to manage the family solo. Young women looking ahead to their future and desiring a family someday have always seen big obstacles to overcome for these types of careers. And there were no workplace structures that would help them make those career choices work.

As a woman's biology and associated social constraints fashion her view of career options, so do a man's fashion his. A young man looks ahead and sees wide open opportunities, but always with the understanding burden that he will always work. Whether he marries and has a family or remains alone, he is expected to work. So, his best career options are those that assure his employability, even if times and job opportunities change. If he wants a family, he must think of career options that make it possible for him to provide support for a family. He must always work and never expect his children's mother to provide the support for the family. A lifetime of career building is his blessing and his burden. Above all, he sees that he must make something of himself. Doctor, lawyer, developer, scientist, professor—these are some of his best options, and they have always required significant sacrifices of family time.

Since the dark ages of 1989, career obstacles faced by both women and men have been either removed or at least softened. As expected, more women are moving into the careers they use to avoid, such as doctor, lawyer, and developer. The Tech workplace, especially, now offers more options for balancing family and work. And men are now freer to participate in those options. Taking time for family is no longer frowned upon or viewed as lack of commitment to the job. As well, men can stay home with the kids and count it as making something of themselves.

This is all great progress, but when it comes to the integration of family life and career, I think there is still a fundamental change that should be made. Fewer people choose to have a family these days, but there will always be those who do. Some make a success of balancing two-household careers with family, while others do not. Wouldn't it be a better future if that success rate were improved? Isn't there great social value in restructuring career paths to accommodate the interruptions needed for optimal family success? I develop this idea in AP 1 on page 227 and offer a rationale for some specific changes.

CH 4. The Lords of Tech

By late 1993 things were happening in Tech that left IBM clinging to the dull edge of Big Iron, still convinced that nothing could ever top it. I wanted no more of their projects. It was sad to see the Golden Age disappear in my rear-view mirror, but I was interested in what I saw on the speedway ahead. The days of interacting with a computer through a command line on the screen were coming to an end. Powerful workstations were starting to run really cool windowed environments with colorful graphics. Icons and mouse driven user interactions were all the rage. Drag it and drop it, honey! A fast, shiny, new Tech car was roaring out of pit lane, and savvy Cleanup Crews wanted a ride.

Tech writers, in particular, wanted to write about these new graphical, windowed applications, and we wanted to use the latest and greatest documentation software to do it. Gone were the days of writing a user guide by typing text formatting tags, line by numbered line, on an IBM 3270 terminal green screen. That was old style. BookMaster was a burnout. Now, we tech writers had FrameMaker, the incredible new WYSIWIG desktop publishing tool, and there would be no going back. Of course, that meant leaving those dead-end IBM projects behind. No problem.

Many of the best, non-mainframe geeks had already left IBM to do the new Tech at innovative, cutting-edge companies that were just coming on the scene. Meanwhile, out in the wild, a generation of developers had matriculated from the newly-formed computer science departments at top universities. They were thought of as the few, the smart, the Tech elite, and they were a hot commodity. All manner of development jobs were now available in the new Tech companies. Some of these companies were very small, starting in a guy's garage or in some storefront that served as a small office. Others were big companies like HP, Sun, SAS, and DEC. They were the ones making the powerful new workstations with UNIX-based operating systems and graphical user interfaces (GUIs). Using those workstations as their development platform, the new software companies were making exciting applications that focused on improving the way people worked in business: finance, insurance, pharmaceuticals, and manufacturing. And these new applications ran fast and processed a lot of data, too. Most exciting of all, they ran on desktop computers. No more green screen computing. Excuse me, IBM, but who really needs that mainframe?

Tech developers were no longer working on programs that just crunch numbers on a mainframe hidden from view in its specially designed cold room—input and output. Now, software development was about helping business employees use a desktop computer to do their daily work faster and better, so that the business could increase profits while maximizing efficiency. The new Tech was about business software that provided workflow processing, production life cycle monitoring and reporting, tracking and analysis of worker efficiency, and replacing paper with online data throughout a business. WOW! There were going to be lots of new Tech projects needing fancy user documentation, and I couldn't wait to make my mark.

All of this wondrous new Tech revolved around the new breed of developer. These guys were into the latest and greatest programming languages, like C++ and Java, that powered the dazzling new GUI business applications. But they also saw that even more powerful coding languages were appearing on the horizon. They reveled in such an exciting, fast moving development environment. It promised greater and more spectacular software as time went on, and only *they* would be ready and able to do design that software. They were few, they were very valuable, and they knew it. Their expertise in the current Tech and their hunger for mastering the fast-approaching new Tech made them a tremendous investment opportunity. A software company's return on investment (ROI) for employing a gang of good developers was like printing money.

It wasn't just clear; it was crystal clear to these up and coming software companies that the success of every one of their software products would depend completely on the ability of their developers to write the code for them. The good developers, and especially the great developers, were in short supply, so the competition for them was fierce. No surprise, then, that when companies landed good developers, they didn't want to lose them. High salaries, great benefits, strong raises, dynamic project work, freedom to innovate, promises of promotion, and the Golden Handcuffs—signing bonuses and stock options—were all put in play to hire and keep the new, hotshot developers.

A software company's culture, in turn, couldn't help but reflect this enormous investment. From the early 90's to the early 2000's, these guys were treated like *prima donnas*. They were young and full of themselves. They occupied a level of importance and authority above the Early Geeks I had come to revere at IBM, and their attitudes reflected it. Not only could these new guys say and do just about anything in a project meeting without suffering consequences, they could also make a software project stop, start, reverse, or go in a different direction altogether. The company culture, which we all had to absorb, was "the developers are always right, and everyone else is expected to defer to them." Yes, they definitely were something special. They were the Lords of Tech.

4.1 The vision of new Tech vs. the reality

The new Tech was changing the face of the everyday workplace. Banks, insurance companies, manufacturing quality control departments—they all wanted GUI software for their employees to use that would eliminate paper processing, and improve the reliability and traceability of their daily workflow. And they were willing to pay for it. The sales and marketing (S&M) people in software companies were selling these potential business customers a compelling vision of new Tech: a virtually paperless workflow that would lead your employees through their daily tasks, prevent them from making errors, and track their work. No more mistakes from moving paper around, no more confusion of what to do when in a process. No more lack of accountability. Individual employee work records would be created automatically and analyzed. Reports of processing throughput would be easy to generate. Your productivity would increase while your losses decreased.

And the vision went further: If you purchase our software product, we will document your specific workflow requirements and customize our software to meet them. We will give you a specification document that clearly states how our software will meet your requirements. When the software is ready, we will install it for you and train your employees how to use it. If you need new hardware, we'll provide that as well. Think of your ROI! With this deal you win, and win big! Remember, your competitors are already moving to the new Tech. You don't want to be left behind. We can help you be first to get there!

Business customers with storage rooms full of paper documents and cumbersome paper-shuffling processes found this vision irresistible. And there were enough of these business customers out there for both large and small new Tech companies to make great profits. So it was that these customers joyfully threw many millions of dollars at new Tech's S&M people. "Yes, yes, we want your vision, we need it, we have to have it!!" the businesses clamored.

And so began the awesome rise of Tech. The Lords were entrusted with the vision, but the reality they had to deal with frequently turned into an impossible mission. Here's how that happened.

Once their business customer signed the contract, the S&M people got to work. They had many meetings with the customer's executive team, during which lots of food and drink were had and the customer's requirements were collected and written down. The S&M people then delivered their requirements document to the lead software development manager in their Tech company. As part of this turnover, the S&M people and the development manager would create a new software project with a budget, a timeline, and a name—typically something the S&M people thought was catchy. The manager would then gather some hand-picked developers together to form a project team and announce the new project, emphasizing all the money it would bring in. To kick off the new project officially, he would hand over the S&M requirements document and tell his Lords to write a document that specified the details of the software they would create to satisfy the requirements.

Hahahahaha!

The Lords of Tech would read the S&M requirements and laugh. They certainly had reason to laugh. The requirements were apples, and the developers were being told to match them up with oranges. It made no sense, and they saw immediately that it couldn't be done. Nonetheless, even though they'd been handed an impossible task, they were much too confident to have a meltdown. They just laughed and moved on. They had to decide how to handle this task in a way that ended up with a successful project that delivered working code acceptable to the customer. It wouldn't be easy, but they had some tricks up their sleeves, and they had the power. I was part of a few of these trick projects, and I would gossip about them with my cleanup crew colleagues at other companies who had very similar experiences. We all became familiar with the tricks, and most of us gradually figured out how to adapt to them. Sometimes the tricks worked, sometimes they didn't. But before I describe those crazy days, I want to explain what had gone so wrong in the requirements gathering process to make the Lords of Tech resort to trickery.

Every software project needs requirements. Where do these requirements come from? The S&M people are expected to get requirements from the customer (typically, the CEO), and the CEO is expected to get requirements from the company's management team. The requirements that the developers need must come from the target users for the software. Unfortunately, these 3 sets of requirements are not the same. They may have some overlap, but for the most part, each set is different. For example, the CEO has high-level concerns, such as the need for solid ROI of the new software, how change management will be supported so the business doesn't suffer during the move to the new software, and how the project's progress will be tracked and tied to its costs.

Those are important high-level requirements, but not what the developers need. The S&M people want to give the Lords something more specific. They believe that capturing business processes for the developers is the way to go. They ask the customer for details about how they want to do document processing, reporting, analysis of data, customer support procedures, recording of employee work, volume of daily workloads, and security needs. During the meetings for requirements gathering, the S&M people and the CEO each strive to get their requirements questions asked and answered. The S&M people tend to assent to the CEO's requirements without a fight and are prepared with detailed PowerPoint presentations that address them. Once the CEO is satisfied that the company's high-level needs are captured, the S&M people work on defining the business process requirements.

To define these for the S&M people, the CEO turns to a selected group of the company's management personnel (typically high-level management) for help. From the CEO's perspective, for example, the directors and vice-presidents of the different divisions and departments in the company know the details of day to day operations. Therefore, the CEO charges them with answering the business process questions. And what do these company managers know about those processes?

They know about their department budgets and bottom lines. They know the mission and goals set for their departments. They know the number of resources (employees) in their departments and the monthly volume of work. They know the business policies in place for managing various situations that occur in the workplace. They know the current successes and failures in their departments and the improvements they want to make. But if these high-level managers were suddenly put in one of the employee groups in their departments who do the actual work of the company, then told to perform the daily tasks of that employee, they would be clueless.

They would be clueless about the specific problems the employee faces in completing tasks, what slows down the completion of tasks, and what would make doing those tasks easier and faster. This ignorance is not a flaw but a natural result of the level at which these managers work. There are managers under them and likely supervisors under those managers who know the details of daily tasks. But they do not.

So what? The S&M people don't really care about those low-level details about the specific tasks of the users and how they perform them. If the company processes insurance claims, for example, the S&M people want to know the volume and the workflow of a claim through the company departments. They don't care what an actual claim processing form looks like or what issues the claims processors currently have working with it. That's for the developers to worry about. That's too much detail to try to cover in the requirements meetings. In any case, the high-level managers speaking for the company don't actually have that information. Whatever they do provide about daily operations is sketchy and general in nature. They may say that they'd like to improve the format of their claim form, but they don't have any further specifics. That's for the developers to figure out. With some mutual hand waving, everyone agrees that additional details can be obtained as needed at a later time, after contracts are signed and the project is underway.

When the S&M people kick off the project and deliver the requirements document to the developers, the desperate Lords page through it frantically, finding only general information about the business space of the customer, their rating in that space, yearly revenues, number of employees, business model, geographical locations, any current technology they have, the statistics measuring the business improvements the customer wants to see in productivity, and a timeline for completing the project. The document also directly quotes the information about the company's departments and their internal business processes and workflow, as provided by the company's select group of executives and high-level managers. As the doomed Lords read through the requirements document, they find little to nothing that describes a concrete piece of functionality or a set of detailed user tasks that the customer needs the software to implement.

To write code, they need to see a requirement like "when a claim comes into our call center, the operator first needs to be able to search a database for all information we have on the customer and have the information displayed on the operator's screen using a specific format, which is provided on the next page." Now that's a requirement that a developer can implement with some code.

Another example of useful requirements would be "the claim form that the operator fills out for the caller should have the fields and formats as shown on the next page. We also need a way to prevent the operator from creating a duplicate claim. And the operator needs to see information that will allow differentiation between customers who have the same last name." Such requirements provide the level of detail that a developer needs for designing and writing code.

A useful business process workflow requirement example would be "within 24 hours of its filing by the call center operator, a claim form must be emailed to an underwriter in the U&W department for approval. If the claim has not been moved to the processing queue of the underwriter within 3 business days, it must be sent to the department manager with a required read receipt and warning flag." A requirement like that may well need more specific information from the customer before it can be coded, but developers can definitely start designing some of the functions they will need to code it.

But alas! requirements like these are nowhere to be found in the document that the Lords receive. And they can't design and write code for the requirements that are actually in the document. How can a piece of code be written that "increases productivity and improves efficiency"? No wonder they laugh. At the same time, though, the Lords don't go back to their manager and the S&M people and throw up their hands. They don't complain that the document doesn't contain the correct kind of information and is, therefore, useless to them. They don't say "Look, we need specific, detailed information about the roles of the target users and how these users do their tasks at this customer's company, in day by day specific detail. We also need an analysis of the way their day by day workflow should specifically be modified to make their tasks easier and faster to accomplish. When you get that detail for us, then we can specify the code that we will write to implement those requirements." Noooooo, the Lords of Tech don't say that.

They don't say that because the last thing the Lords want to do is relinquish their control of the project to others. These others would have the authority to define the specific requirements that they, the developers, would then be forced to implement. The developers would still be responsible for delivering working code according to the project schedule, but they would have no control over changing, ignoring, or redefining the requirements in order to deliver working code on time.

Even worse, who in their company actually knew how to get the detailed requirements from the customer that the developers could use? What process would ensure that only what was useful was captured and irrelevant details were ignored? Getting confusing details, or the wrong details, or requirements that couldn't be implemented in the allotted timeframe would be no better than working from the useless S&M document. Certainly, the S&M people were not capable of gathering detailed requirements. Their focus was on the customer executive who signs the contract. They had little to no interest in what the company's users do or the input they might have. Furthermore, the S&M people didn't have enough Tech knowledge to be trusted with knowing what was important and what was irrelevant. The Lords concluded that, in fact, they themselves are the only people capable of getting the right set of detailed requirements, and no Lord was interested in spending time talking to users or observing day to day business processes. That Lord would be watch-gazing every 10 minutes!

Clearly and above all, the Lords never want to be in a position where they cannot control the progress of a project. It's just too risky. In my experience, when the project reality looked like an impossible mission, the Lords rarely gave up on it. Instead, they would decide to enjoy the project kickoff meeting and then figure out later how to handle the project and its ridiculous requirements document. Often, the plan the Lords came up with resorted to trickery of some sort. But no matter what, they kept things under their control. They would work hard for a successful outcome, but if the worst should happen—a suspended, failed, or restructured project—they would make sure to avoid any blame. In any case, the reality of projects based on inadequate requirements rarely matched the vision sold to the customer.

By the way, if you are wondering whether the problem of confusing and poorly conceived requirements gathering still exists in software projects, the answer is "Yes!" As I write this in January of 2019, companies are still trying to address the problem by hiring someone who can turn high-level business requirements into something a developer can use. For example, a recruiter sent me a Business Analyst contract job description that lists, among others, the following skills for the position:

- Creating documentation that reflects business expert input and is suitable for use by developers
- Facilitate and accelerate the interpretation of business stakeholder input into developer ready documentation

For your reference, AP 2 on page 233 contains the full job description. As I discuss later in this book, Business Analyst is not the role that can solve the requirements problem. But at least we can say that the problem is now recognized, even if it isn't solved.

4.2 *Tricks of the trade*

In the sections below, I recount some of the tricks used by developers on the projects I worked on during the 90s and early 2000s. In cases where the trickery did not affect my work, I did my job and hoped for the best. In other cases, where I happened to be a target of the trickery, I fought back. The Lords always won, and I knew they would. Still, it made me feel better to stand up to them, and there was always another interesting consulting job out there to chase, if I wanted it. I usually did.

Trick 1: Just cowboy it!

I mentioned earlier that I worked on Tech projects that targeted a variety of industries, including insurance, banking, telecommunications, pharmaceuticals, and retail. I found that the senior developers on these projects usually had a strong understanding of the domain of their customer base. On one telecommunications software project I was part of, the primary developers had a good understanding of the telecommunications business the customer was in and the overall functionality that the new software needed to provide. These developers did not have the details about this particular customer's day to day operations or the specific task flows of its target users. But they knew, in general, the typical tasks performed by telecommunications users in these kinds of telecommunications companies.

When the S&M people delivered their high-level, and basically useless, requirements document, the developers on this project asked for clarification and elaboration on a few points. Most of the responses they received were vague, which was no surprise. After that, the Lords decided that they would design the software according to what they thought the customer needed. And to make it fun, they would code it up with the latest-coolest-most-interesting Tech available. They decided to use Java and implement the application in a browser through downloadable jar files. The GUI would have all the new whizbang user-interaction widgets: drag and drop, slider bars, radio buttons, multi-selection lists. Yesiree! and Yee-Ha! the developers would cowboy this project and convince the team that the customer will love it.

Once all this was decided among themselves, the Lords wrote a vaguely-worded functional specification document stating that they would code functions to satisfy the customer's high-level requirements. They gave examples of the "expected" features that would be available in the software, for example, dynamic network monitoring of data usage, data traffic flow monitoring, network error detection, configuration of monitoring parameters, and report generation. To a large extent, they simply repeated what the customer had asked for, adding enough Tech jargon to dazzle and impress.

The customer accepted this specification document, even though it gave no idea of how the features would be presented in the software, or how users would perform their tasks with these features and improve efficiency because of them. Apparently, the customer assumed that these features would deliver the efficiency and productivity that the CEO required, even though the developers' software specification document provided no basis for this assumption. From what I heard during project meetings that were attended by some of the customer's executives, it was obvious that these execs believed the Lords knew what they were doing. And why shouldn't they? After all, these developers were top notch. And everyone says Java is the future!

The developers worked on the code for several months. First they did the design work to identify the discrete functional pieces that would make up the software they had decided to write. Next, they parceled those pieces out among themselves as coding assignments for the project. After thinking about it carefully during these months, they were fairly certain that the project timeline was too tight to include all of the "expected" features they had committed to in their functional specification. (They made no secret of this. You could hear them talking about it in the break room.) Nevertheless, they chose to put all the features in the schedule anyway and worry about it later.

When I saw the schedule, I thought to myself, "Wait a minute—this plan seems a bit too nonchalant. Why start out with a development schedule that they didn't think they could make?" Early in my career on the Cleanup Crew, I found this fictional scheduling very disruptive to my own task planning. But as I came to understand what "schedule" really means to a developer, I was OK with it and adapted. It actually made some sense. Here's what I learned about the project schedule in my time with the Lords.

Before the days of the Agile Software[11] movement, the software project development schedule was a strange animal. It followed a Waterfall[12] project development method. In theory, a Waterfall project followed a straightforward, static task schedule. But in reality, the Waterfall schedule for a software project almost always morphed into a complex of constantly shifting tasks. The reason for this complexity goes back to my earlier discussion of elegant code and the creative aspect of coding.

Every software project contains new features that have never been implemented before, or features that have been implemented before but, for the particular project at hand, some aspects of the previous implementation must be rethought. To manage this, the developers would identify these more difficult implementations at the start of a project and discuss possible coding strategies and the expected time needed to complete them. Because they strive for elegant code, which requires creative thinking and trial and error, at best they can only estimate a minimum amount of time required. They really can't tell you the maximum time they will need. In fact, they have each experienced situations where expected problems get solved quicker than originally thought, as well as situations where coding tasks that did not initially present any problems later turned into a schedule buster.

Just as we can't expect an artist to provide a detailed schedule for completing a new painting, so also we can't expect developers to adhere to the initial Waterfall schedule for a project. We must allow them significant latitude and view the initial project schedule as a strong guideline, rather than a definitive task plan set in stone. Developers will commit to meeting the final delivery date in the Waterfall schedule. What they can't commit to for sure is exactly what features they will be able to deliver on that date. Rather than leave things up in the air, the Lords usually chose to put all the expected features in the Waterfall schedule and let things shake out as they delved into the code. I worked on projects where the Lords were able to deliver on everything in the initial Waterfall schedule. I also worked on projects where they had to scrap features and move them to the next release of the software. Customers came to understand and accept these failures of delivery. Once a customer signed a deal for software, they knew they were making a multi-release commitment and would be getting updated releases over time. Customers tended to ask for everything in the first release, and then settle for a portion of their requests as long as they were promised the remainder in later releases.

As development progresses in the Waterfall project, developers are better able to provide reliable estimates for what they can actually complete and when they can complete it. That's why the initial project schedule contains everything, while the mid-project and final project schedules usually contain fewer features, and often some new features. With this in mind, careful schedule management is critical for communicating the real progress and direction of the project to the Cleanup Crew. The schedule can easily become a cause of failure if team members are working toward features that will never be coded.

As the "cowboy it!" Waterfall project progressed, the time came for the customer to get a demo of the state of the new network monitoring software. The developers were ready for this presentation, and the entire project team and S&M guys attended. Their demo contained a roughly drawn user interface that displayed several buttons. When selected, these buttons displayed the features they planned to demo to the customer. Of course, the features themselves were still in their initial draft implementations, but the customer had been made aware of this. To their credit, the developers had their demo hooked up to a real network. And their demo software was the actual code they were working on, albeit in a rough state and not yet complete. They weren't trying to hide anything from the customer. (This is in contrast to the "smoke and mirrors" type of demo that I saw on several projects. Such a demo is faked up, because the actual software isn't ready yet, so the displays, network connections, and data are created to appear as though they are real under highly restricted scenarios.)

The customer's reaction to the demo was not entirely positive. While the features that were demoed looked promising, the customer couldn't help but focus on the rough appearance of the user interface. "Is this what it's going to look like?" was a question that came up several times during the demo. The customer also displayed some confusion about how the features that were demonstrated would contribute to the company's business processes, on a day to day basis. The demo concluded with a modestly positive reaction from the customer, which the S&M people and the development manager tried to encourage with promises of complete satisfaction when the software was finished. I heard declarations like, "The demo is just a rough idea to show that we can implement the features you need. The finished product will have a polished UI and a complete navigation system. You'll love it!"

The customer left trying to smile but couldn't actually manage it. There was worry in that furrowed brow. And why shouldn't there be worry? What the customer needed to see in the demo was something like this:

> *We have implemented a group of features that allows you to complete a critical task for your business. The task we will demo for you now is the monitoring and management of alerts for a transmission problem in one of the system's lines. The workflow we coded for this task is based on the input we received during our requirements gathering phase with the users in your company. The first step we will show now is viewing the alert. Here is how the system displays an alert.*

And so on. With this kind of demo, customers can see an important task that is done every day in their company, and they can see how the company's users will complete the task with the new software. Now customers are on familiar ground and can actually judge if the new software will improve the company's business process in this one instance. The UI in this type of demo can be rough, as long as it is laid out to show how the user will actually move through the task's workflow.

Unfortunately, the failure of the requirements gathering process, followed by the "cowboy it" approach precluded the construction of the preferred kind of demo. Instead, the cowboys had coded some features they thought were important, and then showed them off as individual, discrete, and unconnected operations. They ran their demo with statements like, "When I click here, see? It does this. Now if I click here, see? It does this." They never showed a step-by-step movement through an important task. No wonder the customer couldn't find much meaning in the cowboy demo and left with a demand for another demo when the user interface was ready. User interface? Now what will the Lords do with that? At this point in the project, they don't even care about the user interface. The UI is something they cobble together when all the underlying features are coded.

The S&M people tell the development manager that the customer needs to see a fancy UI. The software needs some lipstick, by god! OK, so the development manager assigns their least capable guy to put the UI together. They can't afford to waste a really good developer on something as boring and trivial as the UI. They have features to code, after all. So, they tell the UI guy to get some eye candy graphics from the developer down the hall, who plays around with graphic design as a hobby. The UI guy is told to use these graphics to mock up snazzy screens for using the software. In addition, the UI guy is told to use some eye candy for navigating through the features. "Which features," the UI guy asks? They tell him, "Just go by the ones on the current schedule." No big deal. It's a simple task—put some fancy widgets on a screen and hook them up to the features. It's child's play compared to the underlying feature coding the Lords still have to complete.

The UI guy gets started by reviewing a competitor's software to see what he can copy without getting into trouble. He also tries to get ideas from some of the cool Java applications that are popular among developers. He's in a rush, so he fastens on the first couple of UI designs he can copy and reimplements them with the graphics he's been given. So now he has the basic design that will be on every screen and he's got the fancy widgets: drag and drop, radio buttons, and slider bars. The next step is to figure out a navigation system for the features. This step comes down to deciding where to put the menu system. On this particular project, the primary eye candy the UI came up with was a graphic of a large, black TV remote control. His idea was that the buttons on this remote-control graphic would be the menu system. Each main feature would be assigned its own button. He shifted this remote-control menu graphic to the side of the screen. It looked like pretty good eye candy after he mocked it up, except for one problem. There were more features than there were remote control buttons. What to do?

He first tried making all the remote-control buttons smaller, then adding more buttons, but the result was butt-ugly. So, he figured he would design a way to access additional features after one of the remote-control buttons was selected. Ok, now how should these additional features be displayed? Should they be in a list in a dialog box? Should the list be in a menu across the top or vertically along the middle of the screen? He had to make these decisions and make them quickly. But no worries—he came up with a full set of working screens in the required timeframe.

Yee-hah and Yippee, all those durn features were now accessible through the navigation system. He was done! Except for one thing—much of the navigation made no sense. Features were not arranged in any intelligible order, making it impossible to predict where they would be and remember where they were after you found them. Even the Lords were confused when they reviewed the new

UI. Well, they decided, this isn't our problem. Let's just turf this dang old confusing navigation problem to the Cleanup Crew. The developers really didn't care if the navigation was confusing, as long as the features were available somewhere in the UI. It was the user's job to read the documentation and figure out how to use the UI. And it was the Cleanup Crew's job to write the documentation. Time for them to earn their money.

OK, so now it was my turn! For this project I was the Team Lead for the documentation, and our team was loaded with senior-level writers. By this time we had learned what the software's features did, and we had done our best to figure out how these features related to one another. We had also spent a lot of time scraping together our best guess of the target users' daily tasks. After all, our job was to explain to the users how to work with these features to complete their daily tasks. Of course, we only had the meager information gleaned from the various requirements documents and the developers' terse explanations to go by. We didn't actually know how the real users worked in their company. We could only hope that our guesses about their daily tasks and the features relevant to them were what the users needed. In any case, we put together the information we had and documented how to use the software features to complete what we hoped were the user's daily tasks.

At this point in the project, we were working on drafts of our user guide and product introduction. There were big gaps in our documents because we had to wait to see the UI to provide detailed instructions for users. Unlike the Lords, we cared about helping users and were committed to providing clear and correct instructions for them. When we heard that a UI had been coded, we started reviewing it right way. It didn't take long for us to figure out the issues with the navigation. It was so confusing that we didn't think it made sense to try and write a user's guide. Instead, we came up with a reorganization plan for the navigation that would make sense to the user.

We knew that we couldn't change the basic layout of the remote-control menu buttons. So, we focused on the access to the additional features that could not fit on the remote control and were located on submenus. Most of the changes in our plan were minor—-moving menu items from one submenu to a different submenu and changing the names of some submenu items to improve clarity and meaning. However, there were a few major UI changes that were also needed. The location of the submenu on some screens needed to be reworked. In addition, some navigation items were missing on some of the screens, such as a Back option and a Return To Main Screen option. We knew these major changes would not be welcomed by the UI developer, but we felt strongly that they were needed to solve the navigation confusion.

It was my job to sell our plan to development. I started by meeting with the development manager to discuss the confusing navigation. He didn't disagree with me. I told him my team could provide a plan for fixing the confusion, but it would involve some screen changes. He said we should present our plan at the next project meeting, and development would consider it. "Great!" I said. "We're happy to take on this navigation confusion problem! We have a good grasp of how the features are supposed to work together, so we are in a great position to organize the navigation so it makes sense to the users. And the sooner the UI is finalized with proper navigation, the sooner we can finalize our documentation. We'll be ready to show you our plan at the next meeting!"

When the meeting came, I was ready to present our solution. I showed our new screen mockups and explained how our changes solved the confusion in the navigation and helped users. The presentation went well, and the development manager thanked me and the team for our great work. On behalf of his team, he agreed to get our changes into the new UI by the next project milestone date in the schedule. He said he would notify us when the changes were implemented so we could review and approve them. What a great meeting that was with the Lords! I couldn't have gone better. Our team had done well, and our effort was appreciated. We dug into our documentation, basing our user's guide instructions on the new UI changes development was going to make.

The next milestone came, and the new version of the software was made available to the Cleanup Crew's testers. Our tech writing team never heard a peep from the development manager. Undaunted, we marched into the test area and took a careful look at the new UI. A few of our minor changes were there, and that was it. Our major changes, which were the most important, were not implemented, and most of the minor changes were missing because they were tied to the major changes. We looked at each other in shock. "What's going on?" we all thought.

That very afternoon I was scheduled to attend the regular meeting of the project team leads, the development manager, and the development manager's boss. I was livid, but everyone else was congratulatory because "the project milestone had been met." I interrupted their clubby geniality by declaring that the milestone had not been met because the navigation UI changes, previously agreed to by development, had not been made. The development manager looked at me and calmly declared that they did make a few of our changes, but he had never promised to make all of them. I looked him in the eye and said "That's a lie." He chuckled. He said they just didn't have time to make all of those lipstick changes, and the UI worked fine as is. Crickets, crickets, crickets in the room. Hello y'all, remember that meeting where he said they would do it? Anyone home? Alrighty then. The Lord had spoken and was not to be challenged. The meeting continued without me. I was out the door. A few days later my manager told me I had displayed a bad attitude at the meeting. The development manager had complained to him about me. Uh-huh.

The cowboy Lords never did make those UI changes. They delivered the software that they had decided to make, and the customer accepted it. We tech writers tried our best to explain how to use the UI to complete a task, but our explanations were convoluted, at best. A few months later, the company began layoffs. I left for another, better project down the street. About a year later, I heard some gossip that the company was failing and hoping to be bought. Eventually it was bought out, after most of its employees were gone. Now it is no more.

Was this project a success? Did the "cowboy it" trick work? All I can say is that whoever had to use that software was going to be in a world of hurt dealing with its UI. Yes, there were drag and drop widgets, there were slider bars and radio buttons. But fancy widgets couldn't make up for the confusing maze of screens that those unfortunate users would have to find their way through. The Lords, of course, would be enjoying a bonus and moving on to their next Tech job.

First there was the vision, and then there was the reality.

Trick 2: Investigate *ad infinitum*

The project where the Lords used this trick was at a large, national bank—one that has since withered and died. But during the time I was there, it was rich and fat. It had a big, internal software development team composed of both full-time employees and contractors. It also had a small group of fancy Microsoft consultants in residence, and they weren't cheap. From what I could see, they mostly went on coffee breaks at the in-house Starbucks on the lower level. They always seemed happy.

By the time I arrived for this gig, the high-profile project I was to join was in its second year. Much money had already been spent on it. Many valuable employees had been taken off of other projects and dedicated to this one. Yet, after all this time and effort, no software had been delivered, and none was in development. Needless to say, upper management was not happy.

The project's mission was to create software to replace the outdated software currently used for supporting personal banking customers. The current software was used by the call center support department, which had hundreds of operators. The outdated software these operators used for servicing incoming calls had a primitive, grayscale window user interface. This UI provided slow and complicated access to the customer data in the bank's mainframe database. Although it supported a mouse, many of the operator's interactions had to be done with the keystroke combinations demanded by the dinosaur mainframe. Operators struggled with this plodding, clunky display as they tried first to verify the caller's identity and then find that customer's account information. Once that lengthy exercise was complete, the operator could move on to addressing the customer's concern, which required more mainframe searches that took even longer. "Please hold Ma'am while I wait for my computer. I'm trying to pull up that screen now. Our computers are running very slow today, please bear with me." Not a good look for the bank, especially when its competitors were already updating their call centers and advertising fast service with a customer focus.

The original vision for the project's new Tech was a browser-based application developed with, of course, the latest and greatest new programming tools. In this case, it was Microsoft's .ASP, Visual Studio, and database middleware. With the new software application, data would be easier to get and faster to display. Searching for customer account data would be user-friendly and lightning fast so operators could handle more calls and provide better service at the same time. But that's not all. The display would be colorful and fun to use. The windowed UI would be bigger and display more rows of data, and it would support multi-tasking of more than one customer at a time! Efficiency and productivity would be enhanced while improving customer satisfaction. "Yes, yes, we want it, we need it, we have to have it," said all the bank's high-level execs. The vision was definitely what the bank needed, but after 2 years of project work and lots of spent money, the new software was still merely a glint in the eye of the VP of Development.

What in the world had the project team been up to? The true history of the project was revealed during my project orientation. I had to read between the lines, but by this time, I was good at it. At the project's start, the developers had been given the typical useless requirements from both the personal banking managers and also the call center's upper level managers. These requirements included faster searching, multi-tasking of calls, bigger screens, more rows of data, user-friendly displays—all the things that the executives believed would improve call center operator efficiency and productivity.

But these requirements did not provide a deep dive into the current daily tasks and workflow of the call center operators, with details about the issues and problems they faced with their current software. No surprise then, that after reading these requirements, the developers had no clue about the day-to-day task flows in the call centers and, thus, no clue about how features like "multi-tasking of calls" needed to work or what the UI needed to look like to make sense to the operators. Just as we saw in the previous section, the requirements document given to the developers didn't have the information they needed to write a detailed specification of the software they would create. That was bad, but with this project, the Lords had an even worse problem. The worse problem was that, requirements aside, they simply couldn't create the software as envisioned. The Lords had realized early on in the project that the vision was fatally flawed. Here's why.

The bank kept all of its customer data in a "legacy database" (yes, we're talking about the dinosaur mainframe). The bank—like all banks—never wanted to update and transfer this data to a database with new Tech capabilities for one reason and one reason only: SECURITY. The mainframe was a nasty old beast, but by god, no one could break into its database! And, by and large, they were right about that. Customer data was their gold, and they could never put it at risk with new Tech. Not no how, not no way. Who knew for sure that the new Tech could be trusted?

OK, so given that data has to stay in the mainframe database, the architectural design for the project was to use new Tech middleware[13] software running on lightning fast, new Tech hardware. This middleware would be configured to retrieve data out of the mainframe and put data back into the mainframe. The operators would use their fast, friendly new Tech desktop software to talk to the middleware. The middleware would then perform the searches on the mainframe and send the data back to the operators' desktops. This method of searching the mainframe would never expose the mainframe to the less secure, new Tech software on the operator's desktop. This middleware would be lightning fast, and it would have high security equivalent to the mainframe. It would send and receive data between the mainframe and the operator's desktop so much faster than the old operator software that searching and data retrieval would be greatly improved. Boom! We're done!

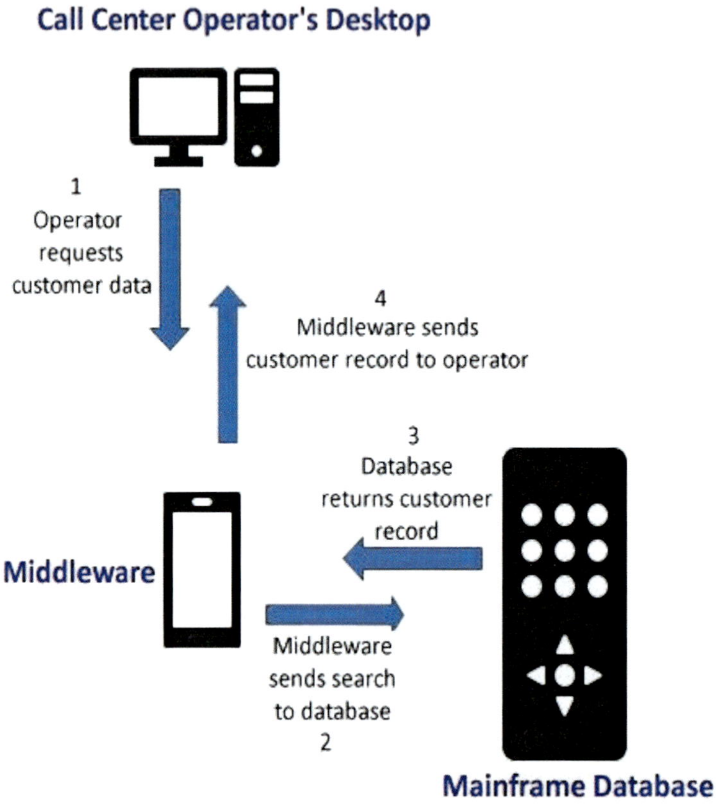

It all looked good on paper. The developers did think the middleware could provide improved data speeds. But they soon discovered that the speed they needed was conditional on the structure of the data in the mainframe database. And, unfortunately, this was a problem. The developers knew that to improve search speed at the operator's desktop, the new software would need more than speedy input and output. They knew the middleware software would also need to provide all of the relevant information for a customer in a single customer record so it could bc delivered to the operator in just one search and retrieve action. The single record could then be displayed on the operator's screen as just one row of data. A single search that returned a complete customer record to the screen would be light years ahead of what operators had now. Currently, they were forced to perform a separate search and review a separate display window for each piece of customer information required to service the customer. The operator had to keep track of all the pieces to get a complete picture of the customer's information. The only way for the new software to overcome this disastrous situation was to deliver the full customer record in one search operation.

Unfortunately, it would be virtually impossible for the middleware to retrieve a complete customer record out of the mainframe's database with just one search. That database was not capable of storing all the pieces of a customer's information as related pieces of data. For example, there was no way to send just one search request to the mainframe database for a customer's account number and get back all of the information related to that customer. Some customer data the mainframe did store as related data, but not all. Multiple search requests would have to be made to get all the customer's data. It was an old style, mainframe database, and the project was stuck with it.

The vision for the project had assumed that an updated, new style database would be available to the middleware. The new style database would have restructured the data so that all a customer's information was tied to one search request. This kind of database would fix the critical search and retrieve issues that the call center operators faced. The middleware would be lightning fast in delivering a single customer record to the operator's desktop, with all the customer's data organized within a single row for the operator to assimilate. But without this new style database, the middleware would have to perform multiple searches in the mainframe database for all the pieces of the customer's data, then restructure the data into one customer record, and then send it to the operator's desktop. Even worse, if the operator made several updates to the customer's record, the middleware would have to perform a separate input/save operation for each update, as opposed to making changes to the data of just one customer record, then saving that record. In short, the middleware would have to perform many more operations than originally thought, and this would slow it down dramatically. The operational speed and efficiency that the bank was promised in the vision would not be possible.

Since the bank insisted on keeping the mainframe database, the developers knew the project's vision could never be realized. They were fairly certain that the middleware would be much slower than hoped for. They expressed their beliefs about the mainframe database problem to their management, but the bank decided to move forward with the project anyway.

"How much slower will it be?" the executives asked. The developers didn't know.

The execs said, "Well then, maybe it will still be better than what we have now, and we have to do something! We can't let our competition get ahead."

Back in their fancy offices, the execs surely had further thoughts, such as, "After all, we hired these Lords at great expense because they could do anything. They were the best. Now it's time for them to spin straw into gold. They can have whatever they need, as long as they fulfill the vision."

And so the Lords became responsible for making the new software that was supposed to meet the requirements they had been given. But it could never be done. What to do? They needed a trick—a trick that would allow them to avoid their problem for a long time. With enough delay, the project might possibly just go away.

The trick they decided on was to investigate *ad infinitum*. They would focus on investigating and refining the requirements until either they came up with a better plan or the project died. The hope was that this trick would allow them to avoid any actual coding, which they thought would be a waste of time. But how could they sell this to the project team?

The developers explained to the project team that they needed the requirements to be expressed with enough specificity and clarity that both the developers and the business management (referred to as "the stakeholders") agreed on exactly what the users needed to do with the new software. To achieve this, the developers proposed a series of regular meetings between them and the stakeholders for the purpose of defining the new set of project requirements. And they proposed to use UML[14], a fashionable systems engineering methodology, for capturing the progress made during these meetings.

Specifically, work groups would be formed, composed of developers and stakeholders. Each group would be responsible for investigating and refining a few of the desired features for the new software, for example, Group A would work on the Search feature and the Customer Information Retrieval Feature. Each group would be tasked with defining the requirements for their features in Use Case[15] documents. One of the developers in each group would be the facilitator and scribe for the Use Case documentation. But all members of a group were expected to decide together how the features would work at the requisite level of detail needed for use cases. Draft use cases produced by these group meetings would be reviewed, modified as needed, and then formally accepted by the group as final. That was the proposal. Did the project manager buy it? He went to the bank execs for guidance.

In the jargon of new Tech, these group meetings were known as JAD sessions. JAD stands for "joint application development", and these sessions were the latest fashion among Tech project teams. The bank execs were excited to have their people working on the cutting edge of new Tech, and they were pleased with the developers' new Tech proposals for running the project. They gave their blessing and budgeted accordingly. What they didn't know is that JAD sessions also provide excellent support for the investigate *ad infinitum* trick that the Lords had decided to use to delay code development. Here's how a JAD session approach slows down a project.

Groups that define software features using JAD sessions and use case documentation inevitably generate 50-75 use cases *per* group. Furthermore, JAD session time is spent reviewing and modifying each use case several times before it's finalized. In this particular project, there were 12 groups. Let's do the math—if each JAD session group generated 50 use cases, we would expect 600 documents to be generated for the project. The groups met for 2 hours each week. In the group of 7 people that I was assigned to, we were lucky to generate one use case per week, and that was only the first draft. Now what kind of time frame should we be looking at for these groups to complete their tasks? It's a scary calculation, and no one on this project looked ahead that far. Instead, they had the kickoff luncheon and set to work.

Before the JAD sessions could begin, the stakeholders needed training, and this would take at least a few weeks, given their busy schedules. They had to be taught enough UML and use case jargon to be able to participate fully in their JAD sessions. After they were appropriately trained, a master calendar for the JAD sessions had to be developed, which also took some time. The weekly 2-hour sessions for the 12 groups were to be run concurrently, but they could not interfere with other, unavoidable work commitments of the group members.

Six months into the project, the training was complete and the meeting schedule was settled. The JAD sessions began. Initial progress was slow, as the stakeholders wrestled with the minutiae they needed to provide for each use case. Most of their preconceived ideas about the new software were based on scenarios in which operators perform an initial task and then use the results of that initial task to perform additional tasks, until the scenario is complete. But the use cases they worked on demanded a much more granular approach. They needed to focus on just that initial task and then subdivide it into the steps and, if needed, substeps, until all possible user actions and system responses were identified. "What do you want the new software to do at each step and substep of the task?" the developers wanted to know. "Well, aren't all these details what you developers are supposed to figure out?" the confused stakeholders would respond. "Aren't we just supposed to identify the task, and then you developers work on the steps for the task?" they asked.

In time, the stakeholders came to understand what they needed to contribute, but still, they never felt they were accomplishing very much. In their imaginings, they had hoped the group's work would end up with pictures, or diagrams, or even crude sketches of what the software would look like to the users. But they were way off the mark. Use cases written with unfamiliar expressions and hard to imagine, low-level action sequences were all they had to show for their efforts. (See AP 3 on page 235 for a use case example that is similar in structure to the ones the stakeholders had to contend with.)

The JAD sessions marched on for about a year, with no code written and dwindling enthusiasm within the groups. Stakeholders saw no end in sight and started to grumble to their management. Eighteen months into the project the Banks's execs saw high spend on the budget, hundreds of use cases documents in varying stages of completeness, and no code. The dreaded "analysis paralysis" had taken hold. But while the stakeholders were discouraged, the developers remained committed. "These use cases are what we need," they said.

That's where things stood when I and 4 other contractors finished our orientation session and joined the project's Cleanup Crew. But as we were leaving the orientation session, the product manager who hired us emphasized that he expected us to help the groups get results faster and get the project moving. He speculated that the project had roughly 6 months to deliver some working code or face being shut down. We figured it wasn't just a coincidence that we had signed on to 6-month contracts, with the possibility of renewal.

As Technical Writers, we were assigned to the JAD sessions and tasked with taking notes and working with the stakeholders to help them define the use cases. As we got to know one another, the stakeholders confided in us, declaring their frustration with the current process. They desperately wanted to see something running on a computer or else just give up. A month into our contracts, the stakeholders told us that they wanted to abandon their JAD sessions and delegate the job to us. The sessions were sucking up too much of their time, and now they had other work commitments that could no longer be put off.

Among the 5 of us, I and another fellow traveler (I'll call her Molly) were OK with this idea. But we had one condition. We needed to understand what the users needed from the new software. We told our stakeholders that we needed to understand how the call center operators did their daily work, and to do that, we needed to observe them. The stakeholders thought this was a great idea, and the very next day, they set us up with headphones at the desks of a few of their seasoned operators.

Molly and I spent 4 days listening to calls and watching how the operators used their software to service their customers. We discussed our questions about what we observed with the operators after each session with them. We came away with written reports of detailed user stories that described the operators' daily workflows, their specific issues with their tasks and the key problems they had with the software they were using. Now we were ready to work on use cases in the JAD sessions and complete them quickly. And we were also ready to shake things up in those sessions. Here's what we did.

During the next 2 weeks of our JAD sessions, we distributed our user reports. The stakeholders, who had announced they were soon leaving the sessions, declared how pleased they were with these reports. They went on to say that these reports demonstrated that Molly and I now had the requisite business knowledge to serve in their place. This meant that we had the full support of our stakeholders to represent the business in the JAD sessions. The developers were not particularly happy with these changes, however. They didn't like that we had established ourselves as sufficiently knowledgeable about users that the developers had to accept what we said. They especially didn't like that the stakeholders had given us their full support. As powerful as these Lords were, they had to treat the stakeholders as equal partners on the project. And now, that meant they had to pretend they cared about what Molly and I said.

We knew this turn of events was our chance to blow up the JAD sessions. And we did. At the last meeting that our stakeholders attended, while everyone was once again drowning in use case minutiae for a desired feature, we interrupted the paralysis with a suggestion, "Let's try a different approach that can help us express what we think this feature should do. Let's try sketching out some screens that show how we think users need to work with this feature."

All heads turned our way. After a few moments of shocked silence, the stakeholders joyfully shouted, "Yes! Let's try that!" The developers went quiet and gave us a challenging look. But we were ready. We had brainstormed all week about this feature in the context of what we had learned about the operators. We came up with several screen sketches that we thought were good candidates for discussion. When the developers asked if anyone had any ideas of where to start, we stood up and said, "Sure, we have some," and we started sketching on the whiteboard. The stakeholders chuckled gleefully. The developers were silent.

Of course, as we foresaw, the sketches were a big hit. Everyone got excited, offering suggestions and constructive criticism. Even the developers were into it. Before the meeting ended, we suggested that at the next meeting Molly and I would come with updated sketches of those we had all worked on, plus new sketches of additional screens related to this feature. Everyone agreed to this and the stakeholders left energized, with hope that the group was finally getting somewhere. But Molly and I had yet to deliver our final blow to the JAD sessions.

The next day we met with our manager. He had heard good things from the stakeholders and congratulated us on the dramatic progress made in the JAD session. Then he asked us what we recommended for keeping things moving forward. We had prepared for this question and hoped he would ask it. We discussed the frustration of the stakeholders and their desire to leave the JAD sessions. We all agreed that losing their participation would threaten the success of the project. We noted that the developers also felt this way, and our manager agreed wholeheartedly. Then, we proposed stopping the JAD sessions altogether for 3 weeks. During this time, Molly and I would prototype a set of screens that would show the main window of the application, including a navigation system leading to the full set of features. In addition, we would provide screens to show the window flow of a customer record search and retrieval, including how the retrieved customer data should be presented in the window. Finally, our screens would show how the main window supported working with multiple customers in the same session, which was one of the more important features the stakeholders had asked for.

Our manager was all ears, so we continued describing our plan. When we finished creating these screens, we would connect them up into an interactive prototype that we could demo to the entire project team. With this demo, the team could see how to perform a few important tasks with the screens, and we would explain why the task flows we designed met the needs of the operators. During the demo, we would gather the team's feedback. Of course, we would only have an initial set of screens that covered only a few important tasks, but this would be enough to decide if the team thought we were on the right track. Oh boy! Our manager was excited, nervous, and relieved all at the same time. He fully supported our idea, hoped we could actually do it, and went off to sell it to the team.

We were asking for a big change to the project's direction, but both the stakeholders and the developers were happy with the idea. From their point of view, the main benefit would be getting a break from the JAD sessions. Truth be told, everyone was tired of them at this point. Having 3 weeks off from them would be a blessing and well worth it, even if Molly and I failed to produce anything reasonable. So, everyone was cool with giving us time to work a miracle.

We dug in, working feverishly for long hours and loving every second of it. We knew we could do it, but from the disinterest the developers showed us every day, it was clear that they believed we would fail. It didn't matter to them one way or the other, though. If we failed, the JAD sessions would likely resume, which was OK. Alternatively, if we failed, maybe the stakeholders would give up and the project would be cancelled. This would be the best outcome, since the failure would be on the contractors' shoulders and not the Lords. No matter. Molly and I just smiled at them and said to each other, "OK, time to knock their socks off." And we did.

Molly was great at figuring out the interactive structures we could use on the windows. I was great at figuring out navigation flows, based on the needs of the operators we had observed and interviewed. Both of us could use Dreamweaver[16] to prototype the screens. We made a storyboard first, then divided up the prototyping tasks and finally put it all back together. It was an HTML interactive prototype, but of course, it was not real code. It was smoke and mirrors to show what we thought the real UI should look like. We made no secret of this, and the team understood what a prototype meant. We also made a PowerPoint show of the screens, since the bank execs always wanted to see a PowerPoint. We even wrote a User Interface Specification document that contained pictures of the screens and descriptions of the task flows across the screens. After three weeks of hard work, we told our manager to schedule the review meeting for our demo. We were ready put on our show.

The demo was a crazy experience, but in a good way. The stakeholders loved the workflow of the screens. They got very excited when we showed the flow of an operator doing a customer search. They could see, from the screens, exactly what the operator would click for each step in the task, what would happen on the screen, and then what the operator what the screen would look like when the task was complete. The prototype was smoke and mirrors, and of course we explained that it was a simulation, but the stakeholders couldn't help themselves. For so long they had hoped to see something working, and this prototype looked so real. It seemed to work like real software, and it did just what the operators needed done. The stakeholders were clapping their hands, and talking out loud, "The call center needs this. We have to have it. The vision has finally come to life!!"

They kept asking to see how the other navigation items on the screen worked, and of course, we had to say we didn't want to design those until we had confirmation that our prototype was what they wanted. They stated that they loved the prototype and wanted us to continue with the design. They told development that they wanted to put our design in the call centers as soon as possible. They wanted to know the next steps that were needed to turn our prototype into real software. Our manager stepped in and told them that the design team would work with development to start coding what we've prototyped. Further, Eva and Molly would continue building out the screens and reviewing them with you until the design was complete. Molly and I jumped in to ask if we could test our prototype with some of the call center operators to get their feedback. Yes, yes, the stakeholders agreed, we'll arrange it—whatever you want!

Molly and I left that meeting as triumphant heroes. A week later, we each received an award from the bank's upper management for superior work on the project. The award came with some money, and it was unusual to give these awards to contractors. But the execs were happy, the stakeholders were happy, the rest of the Cleanup Crew was happy. Finally, at last they had something they could work with—something concrete—a real conceptual model of how this software will work.

Yes, they were happy, but in the halls of development there was a dark, deep silence. In their minds, the Lords must have been thinking, how dare they? Are we Lords now expected to look to *them* for how to proceed? Is someone suggesting that we Lords are to be code monkeys for Molly and Eva? Of course not, they would have said to themselves. That will never happen. Furthermore, this project will never be coded under any circumstances because it cannot be coded. It's impossible, and Lords do not task themselves with such foolishness.

At this point we were starting the 5th month of the project, so we really had only about 6 weeks left to show upper management that the project was worth extending. Molly and I worked feverishly on the remaining screen designs. We were able to create an interactive prototype that included about 90% of the features. We tested it with about 10 operators, recorded their feedback, reviewed it with stakeholders, and made appropriate changes. For their part, development had responded by assigning one, new contract developer to us.

We were supposed to go over our prototype with him so he could start coding the UI of our main screen. Yes, we were given one guy, totally new to the project, who was supposed to work with us to code up the UI all by himself. If he had problems, he was to report back to development. Clearly, the Lords themselves wanted no part of this UI work.

Our guy was slow. He wasn't happy with his assignment. He dithered and fussed, often telling us that the interactions we wanted required "custom code," and he was only authorized to use "out of the box widgets". Sometimes he growled that what we wanted "just couldn't be done". We found his fussing hilarious, because the developers had repeatedly touted the power and capabilities of the hugely expensive development tools they had purchased for the project. These tools were the best of the best, and yet our guy was not allowed to consider innovation or custom coding while using these powerful tools the bank had paid for. Whatever, we at least got the guy to code enough real, working screens to give the first code release for the project to the testing team. This was quite a milestone. Years later, after Molly and I had each had moved on, we recalled this period as one of our most productive work experiences.

During those last 6 weeks of work, the execs and stakeholders were fairly confident that the project would be a go. They continued reviewing our new screen designs and helping us as needed. The JAD sessions had not resumed, and no one complained. While all this was going on, we learned that the Microsoft guys had approached development with some new ideas about solving "the middleware problem." They weren't sure it would work, and it would require buying even more pricey Microsoft hardware and software, but it could turn out to be a totally great solution. This possibility made Molly and me even more hopeful that the project could succeed.

How did the Lords respond to this news? Hmmm, scowl, scowl, they thought about it, then told the Microsoft guys "We'll think about it. Send us some white papers with more details, and let's schedule some meetings with your database guys back in Redmond. Right now, we're busy with our new plan for writing the use cases for the project." WHAT???

Yes, the Lords still wanted their use cases. Their rationale, which they announced at the next project meeting, was that screens were not what the developers needed to write the code. They agreed that what Eva and Molly were doing was fine for getting user feedback, but the screens were just "the lipstick" of the software. The developers still needed use cases before they could code the guts of the software, and fortunately, they told the team, they had discovered a way to get them done much faster and with less effort than the JAD session method. They had decided to employ the latest and greatest software fashion to complete the use cases—the Agile Development Method. But first, they needed training in this new method. And they weren't kiddin' around, Honey. Development management had forked over thousands of dollars for Alistair Cockburn[17], the Agile man himself. He was coming to the bank for two weeks to be their Agile Coach. His training sessions on writing Agile use cases would start soon, and Molly and I were expected to attend, since we could help with the documentation of the use cases.

Well, that was ridiculous and would be a drain on our time, but we would work with it. As long as we could continue our UI design work, we would go along with this use case craziness. After all, we figured, you have to give some to get some. We knew our UI design work was going to make this software a delight for our target users. For us, making software that satisfied its users was worth enduring more use case pain.

The stakeholders, of course, wanted no part of this new Agile stuff. They were done with use cases. Molly and I would have to shoulder that burden. They were focused on a big change that was happening in the bank's top management. A new senior Vice President had just been hired, and the word was that his mission was to cut costs. How would he feel about our project? A lot was going on during those 6 weeks. Everyone had their expectations and hopes for the project, as well as their viewpoints on its value. In any case, the general feeling among the Cleanup Crew was upbeat, and our manager was encouraging. But we were all mistaken.

As the last week of the 6-month project deadline drew nigh, the use case training had finished and the developers were smiling. Why? The bad news leaked out and spread across the cubicles on a Friday morning, before the official announcement. A meeting of the full project team was hastily called for that Friday afternoon. Great work everybody the VP of Development told us. You accomplished so much, and the bank is proud of all of you for your contributions, blah blah blah. We contractors tuned out during the awards and clapping and came back alive when they got to the important stuff. The project would be shut down immediately. Management would be considering next steps. Contractors were asked to stay on for 2 more weeks to tidy up any details. Thanks to everyone and Godspeed.

The Lords had gracious smiles on their faces and, as we dispersed and left the room, they thanked the individual members of the Cleanup Crew for their hard work. Molly was tearful, but I had seen this trick before. Sure, I was disappointed, but I remained unmoved. I was encouraging to Molly. At least we had been able to do the right and proper work for making user-friendly software, I reminded her. That was a lot better than wasting time on use cases and trying to document poorly designed software. I encouraged her to think of it as great experience that will lead to more opportunities to do design work. And I was right, for both of us.

Did the Investigate *ad infinitum* trick succeed? Yes, it did. The Lords were able to drop that project and suffered no blame for its failure. The new SVP eventually decided to outsource the project. I heard later that his expensive outsourcing plan failed miserably, and soon after he left the bank to "pursue other opportunities". The software was never created.

Our manager at the bank offered me a UI design contract in a different area of the bank, for even more money. I took it—after all, that bank had a Starbucks on the lower level. Molly decided to take a gig in a different state, closer to her family. She went on to do great UI design work at another, even bigger national bank, which was no surprise to me. She was a freight train that couldn't be stopped. She had a two-year-old at home, a supportive husband, and was two months pregnant. She wanted a design career and went after it like a tiger. She was managing the superwoman gig pretty well, yet during our bank project days she had confided in me that she wasn't experiencing the happiness she thought would come with it. The stress of absentee motherhood weighed heavily on her. I felt her pain but had little to offer in the way of a solution.

As for the operators in the call center—the users we were supposed to help with our work—well, they kept plugging away with their same old, miserable software, saying to each other, "Yeah, we knew we'd never get that new software. No one really cares what we have to deal with." That's a shameful refrain, and, to my regret, it wasn't the first time my work was associated with it on projects run by the Lords of Tech.

First there was the vision, and then there was the reality.

Trick 3: Prototype and procrastinate

While I was a senior tech writer, I took contracts with 2 software startup companies[18]. I have to admit that the first startup had me fooled to the very end. In my defense, it was only the fourth contract I had worked on as a consultant, so I still had a lot to learn. What a crazy ride that was.

It was 1995. An ex-IBM sales executive had taken a very lucrative severance package from IBM and decided to start his own software company. He knew a couple of developers who were interested in his idea, and they got things rolling. Their first order of business was to identify a target market. They chose the insurance market. They had some knowledge of the insurance claims process and had some business contacts in that industry. There were some big players in that domain and a lot of money available for IT investment. Certainly, the ex-IBM sales exec—I'll call him "Lewis"—had experience with companies at that level and was confident he could successfully pitch some software products to one or more of them.

After some period of brainstorming, Lewis and his head developer—I'll call him "Martin"—came up with an idea for a software product that would automate the claims processing workflow in an insurance company. With this idea in hand, Martin set about gathering a few developers who were interested in moonlighting for the new startup. This small, part-time team crafted a good-looking prototype that demonstrated some basic claims processing workflow. It had drag and drop, it had slider bars, it had fancy selection lists. Of course, it was only a prototype. It was hooked up to a set of dummy data that drove it, and it wasn't usable for anything else. It could not connect to a real database, and its code was not scalable for use in a real insurance claims processing environment. But that was OK. It was a way to show potential customers what they could have if they invested in the development of the actual product. Essentially, the prototype that Martin and his part-timers created was the vision, and Lewis could now go sell that vision to insurance companies.

Like all the other large companies of this era, the insurance companies were desperate to take advantage of the new Tech. And so it didn't take long for Lewis to land a big fish. His new customer was a major national insurer, headquartered in the Northeast. This customer saw the prototype and bought the vision, hook, line, and sinker (now there's a word with some serious foreshadowing).

Lewis worked with the customer on requirements gathering, and, from time to time, Martin was also involved in answering the customer's questions about what the new software could do. The customer focused on the prototype, asking for specific enhancements and additional features to be put into the real software. Lewis agreed to all of these requests. He and Martin created a requirements document that the customer accepted. The budget was defined based on this document, and the customer agreed. A project plan was developed, and things started rolling.

The customer handed over large amounts of money so Lewis could, essentially, outfit a software development company. Lewis started hiring, paying top dollar. Full-time developers were brought on, and many contracts became available for Cleanup Crew personnel. I was one of those contractors. We all started in a small office space, but Lewis had purchased high-end SUN workstations for all of us, interconnected with the latest new Tech networking infrastructure.

Everyone was busy and excited at the start of the project. We tech writers were tasked with both user documentation and internal developer project documentation (e.g., high- and low-level software specifications, configuration manuals, project plans). The customer wanted to be involved with the documentation, reviewing outlines, and then early drafts, so there was no time to waste.

Meanwhile, the new development team began planning the code. The requirements document was typically useless. But they did have the prototype, which the customer loved. So, they focused their attention on the features shown in the prototype, knowing that completely new code would be needed for the real software. In the first few months, they reviewed the requirements document *pro forma*, while concentrating on the features in the prototype. After that, they dug into their investigation and code design phase. As usual, that's when the problems started.

First, they realized that the features shown in the prototype were going to take a lot of time to design and code. Furthermore, they discovered that they would need significant time to design the new Tech backend hardware and software architecture, which was needed to support the features shown in the prototype. Databases, servers, security systems—all of these would need to be defined and in place before the developers could even test any of the new workflow software they were hired to create. The customer had none of this backend architecture in place, and worse, the developers that Lewis hired had little expertise in these backend hardware and software systems. Their expertise was in end-user workstation software, as demoed in the prototype.

In addition to these concerns was the obscurity of the requirements document. As I described in the two previous projects above, this document did not provide the level of information and the detail of information that the developers needed to design specific software functions and screen flows. They took their concerns to Martin, and he went to Lewis. These two realized that the developers' concerns were showstoppers. There was no way Lewis and Martin would be able to deliver the software on time and within budget. What to do?

Lewis actually believed he could sell the customer on providing more money for developing the backend architecture. The extra money would be used to hire another set of developers with expertise in hardware and software backend architecture and security systems design. This new group would start right away, so the backend setup would be ready by the end of the project. But the problem of the workflow software needing more time than the project allowed still remained. Lewis turned to Martin for a solution.

Not to worry. Martin was a Lord, and he had a trick. They would continue refining the prototype and pass it off as the real software. At each project milestone for code delivery, they would demo a new version of the prototype and tell the customer it was the real thing. But they would remind the customer that they had to use dummy data and dummy hardware until the backend architecture was in place. Therefore, the new code could not be installed at the customer site until the backend setup was there.

With this trick, Lewis could pretend to deliver new code on time and demo it, and the customer would have no way of telling that it was still only a prototype. Of course, Martin would have a team working on the real code, and eventually, he would have to deliver it. But this trick might give the Lords enough time to figure out how they could talk the customer into lengthening the project schedule. Prototype and procrastinate.

Lewis was able to sell the customer on the additional new Tech backend architecture plan, and another boatload of money was delivered to his startup company. More developers were hired to work on the backend architecture, and the Cleanup Crew was told that code development on the workflow software would continue in parallel. We tech writers then became responsible for the backend architecture documentation as well, so our ranks of contractors had to increase. Things were hectic, but spirits were still high as the money flowed. We moved offices twice more, each time into more spacious and grander digs. I became a manager of the end-user documentation group, and I focused my team on working closely with the developers to provide feedback on any usability problems they saw in the software. We didn't know we were only looking at a prototype. It was a well-kept secret. The most we were ever told was that the software we were reviewing in the test lab would be easy to transfer as the final deliverable once the backend was up and running.

Like us, the customer was tricked into thinking Martin's incremental prototype code deliverables were the real thing. As customers are wont to do, each deliverable review resulted in more customer requests for changes and additions. To the customer's endless delight, Martin was always amenable to those requests. Martin also accepted input from my team on UI changes that would improve usability. He knew, of course, that modifying the prototype was easy and, as a bonus, Lewis could charge the customer more money for the changes. Yep, things were working out great. Personally, I was extremely satisfied with my role. I felt my team was doing justice to our end users by carefully molding the user interface into something that was, in large part, easy to use.

I was even invited to visit the customer site to observe their claims processors at work, report on the current workflow they used, and identify the problems with it. I accomplished this task with great interest and competence. The users I spoke with became very excited when I described the features our new software would provide. They were relieved that someone understood how they worked and what they needed. They were hopeful that their workflow process would be faster and easier. They expressed concern about learning how to use the new software, and I assured them that our company would be providing training as well as complete user documentation.

I returned to my team and the development team with a detailed list of user needs and ideas about how our new software could accommodate some of them. Everyone was proud of my efforts and pleased that the customer was so receptive to my ideas. My team was further energized to work with the developers on the suggestions I came up with.

After 9 months of this charade, cracks started to appear in the process. The customer was anxious that no software was yet installed and working at their pilot site. They did not want to provide their next cash infusion without some tangible results. Lewis was harder to work with because he was often traveling and unavailable. In fact, he had started to woo another insurance customer on the west coast. Perhaps he was anticipating a project default with our current customer and wanted to get another on board before all hell broke loose.

A few months later, the dam did break, and it was sudden. We had just heard a rumor about the code not being real but didn't believe it. The Friday of that week, we walked into work in the morning and by afternoon, with no explanation, half of us were fired, packed up and escorted out with our box of personal items. It was a beautiful spring day, and I had ridden my motorcycle to work that morning. I had to call my husband to drive to my office because I couldn't fit the box on the back of my bike. Hahahahaha!

The carnage continued at that company. Developers left for greener pastures. Those that stayed were later fired. A few of the Cleanup Crew were kept on for several weeks to tidy up. That's when they found out the code was definitely not real. I had another, even better gig by then. We all met for lunch and laughed about "the prototype," speculating about how they had so thoroughly fooled us.

As for Lewis and Martin's startup company, it sank like a stone. The waters closed over it, and by that summer it seemed as though it had never existed.

Did the trick work? A lot of us in that startup made a lot of money and boosted our resumes. Lewis later went into local politics with some success, while Martin fell off my radar. I heard later that the customer was bought out and absorbed by a larger competitor. But I haven't forgotten those users I met with at the customer site, and how I let them down.

Yes, first there was the vision, then there was the crushing reality.

4.3 Users—Tech's beasts of burden

You've likely noticed the sad thread that runs through my tricked-up project experiences with the Lords of Tech. In the end, they skate, but the user always gets screwed. I don't blame the Lords for using their tricks, though. Most of the time, I could see that their intentions were good. They used their tricks with the hope that, somehow, a trick would give them enough time to figure out a solution to their code problems so they could deliver a good product to their customer.

Clearly, the requirements gathering process at the start of a project often failed to give them what they needed to design and code the right software. Furthermore, the requirements gathering was about the only part of a project that the Lords could not control. The S&M guys took charge of that, often manipulating the process to help sell the vision to the customer. Dealing with the reality was left to the Lords, after the contract was signed and promises were made. For sure, it was a tough setup to walk into. What I finally had to renounce as unacceptable, though, was the attitude of the Lords toward the target users. Let me explain why.

The Lords never focused on satisfying their customer's users. Instead, they focused on satisfying the customer, who was typically represented by members of the company's upper level management team. These managers wanted improved efficiency and productivity, as sold to them in the vision. But these managers didn't know that the Lords didn't care if the customer's users didn't like the new software or had trouble using it. The Lords would make software that *could* improve efficiency and productivity for the customer, and that was as far as they would go. That was how they defined customer satisfaction. It was up to the Cleanup Crew to explain to those pesky users how they should provide the software program with the input it needed to do its job.

This was how the Lords thought, and it was a throwback to their heritage—the ancient IBM mainframe days. The mainframe program was very specific about the input it required to perform an action. The input was very alien—you had to know commands like

```
CARDS
```

and

```
//SYSPRINT DD  SYSOUT=Z
```

The mainframe demanded the right command at the right time and in the right sequence and, as the user, it was **your job** to give it **exactly** what it needed. If you did your job, the mainframe would put out amazing, wonderful, invaluable results for you. Much was demanded of the user, but so long as he be faithful, much would be given unto him. You just had to learn to give the machine what it needed. Basically, you had to learn how to program the machine.

With this mindset, the Lords really didn't worry about users. In particular, they didn't worry about how the user interface looked and whether it made sense. Yes, it would have nice colors and graphics, and pull-down menus, and dialog boxes, and drag and drop functionality. But in fact, the UI was the last piece to be coded, and the flow of the screens would be predetermined by the underlying code that had already been written. The order in which the code processed the steps in a user task turned out to be whatever order made the best coding sense to the developers.

Developers put the UI together by looking at the code, then showing one or more entry fields on a screen at the point in the code's processing where user input was needed. It was then someone else's job to explain to the user what each field meant and exactly what type of input was needed. All the Lords needed to do was show that, when provided with the right input at the right time, the software did its work and improved efficiency and productivity. Boom! They were done.

In a successful project under these terms, the software would be delivered and installed, and the Lords would arrive to give a glorious demo that illustrated the power of the new system. The customer would watch with a combination of excitement and fear. The sequence of screens the Lords flipped through, back and forth and every which way, looked like a strange and treacherous roadmap. A few minutes after the demo was finished, the customer would realize that he couldn't remember the first thing about how to make the new program work. But the S&M people would reassure the customer that a full set of user documentation was included with the software, and customer support would be available free for 90 days. Furthermore, for additional cost, training could be provided to the users.

The underlying message to the customer was clear. The Lords had demonstrated that the software did what it was supposed to do. Documentation describing how to use the software had been provided. Therefore, the burden of using the software successfully was on the user. If your users aren't able to learn how to use our software, well, clearly the failure is theirs. And that's your problem, not ours. Goodbye, and enjoy the glory of your new Tech.

Message received! The customer would charge management with getting the new software working successfully and smoothly across the board. Department heads were told to distribute the user documentation and provide training as needed. Goals and deadlines were set: Here is your training budget and timeline. Everyone is expected to come up to speed without complaints. Make it happen!

Easier said than done. Often, as I've described, the software was unusable. Users were given the documentation and went to training. They worked hard at trying to please the new software. They tried hard to give it the correct input. They knew it was their fault if things didn't work. They knew it was their job to figure out what the software wanted. They struggled every day. The documentation had to be constantly by their side, and god help you if there weren't enough copies to go around. You can have my firstborn, but NEVER TAKE MY USER MANUAL.

No matter how many times they performed a task, they couldn't remember how to do it reliably. Tasks always seemed to have too many steps across too many screens. Worse yet, if they made one mistake in the process, they were screwed. First, they had to stop and figure out what the error message on the screen meant. For example:

Out of memory

Then, they had to try and recover from the error. What to do? They would try what the book says, but most of the time, it didn't work! (Probably one of those last-minute bug fixes that came out after the documentation was printed and, oh well, that procedure won't work anymore.) Most of the time, all they received for their diligence was another error message. For example:

Incorrect address!

My colleagues and I would get frustrated when we had to document such useless messages. To cheer each other up, we used to pass around our favorite list of fake but tragically perceptive haiku error messages[19]. Here are a few of the best:

Out of memory.
We wish to hold the whole sky,
But we never will.

Error messages
cannot completely convey.
We now know shared loss.

Yes, these error messages are jokes, but most of the real messages in software of that era were no better. I recall we tech writers asking development to let us craft the error messages so they made sense to the user. We were refused, since the messages were "part of the code." But even when the messages were isolated into separate text files, developers were reluctant to let us do more than make a few grammar changes. As a result, in addition to a confusing UI, users had to wade through inscrutable error messages.

I really give these dear users a lot of credit. Most of the time, they invented ways to work around the craziness of the UI and get the software to do what they wanted, or at least, what they needed. They often still had to perform some tasks manually, but they kept these glitches hidden from upper management. They never publicly blamed the software for the glitches, but at lunch with their trusted BFF colleagues, they laughed at it until they cried. They knew they weren't the stupid ones. They persevered because they needed the job, and because, for many of them, they weren't going to be beaten by a dumb machine!

These brave users had tricks of their own, too. These tricks are likely familiar to every reader who's used software in the workplace. The infamous "cheat sheet" was one of the most effective tricks. On this secretive paper, duplicated by each user and slyly taped to a pullout in the desk, one would find the way out of the software's most common problems. Usually, the way out was the same—reboot. But the cheat sheet also covered all the gotchas associated with rebooting.

Usually, the sheet helped you out with login problems too, especially with identifying the "right" person to call. That's the nice IT guy who won't report that you were "having problems with the software login again." Finally, the sheet told you what to ignore when the software got pissy and made threats like "All data will be erased. Continue?" And it showed you how to circumvent the software's refusal to complete some task, for example, "Now, when it says the name you provided has already been accessed, and you can't continue, just click the Delete button, then hit Enter. It will take you to a window where you can select the name you need. But if it still doesn't work, ask Joan."

Ah, Joan. She's another of the great tricks the users came up to beat the ridiculous software into submission. She was the resident guru. For some reason, every department had a Joan—the person who thought it was fun to beat the machine. This person delighted in figuring out how to overcome problems when all else failed (documentation, support calls, advice from other users). The resident guru never gave up. She completed her regular tasks, and still found time for testing. She would make a reported problem appear so she could mess around trying to fix it. She was totally open-minded about things to try, because she knew there was no rationality to the way the UI worked anyway. Eventually, the guru would find a way to solve the problem or work around it.

Joan was much beloved by her department. There was always someone at her desk asking for help and usually leaving with the look of a rescued swimmer who had been going under for the last time. Even the manager went to her for help and wasn't ashamed of it. These **Joans** loved helping people too, because they were always hoping to find a new problem that they could solve. It was fun! They made the software look good, even when it was a dog. In a way, that's too bad. They didn't get any credit, and the software remained in place to continue annoying people and making their jobs harder than they ever should have been.

So it was that, in many cases of new Tech's software, those "computer illiterates," those "error-prone" users, took up their burden with deference and gradually overcame the unthinking roadblocks that the UI presented. Sometimes, a measurable increase in efficiency and productivity actually occurred. Sometimes even hard to use software was better than no software at all—as long as you had your cheat sheet and a resident Joan. But many times the productivity statistics were pretty much unchanged.

Then there were the users who had to suffer through several years and iterations of new Tech software that was impossible to deal with and was never going to work right. In these cases, productivity and efficiency decreased. Everyone knew it, but it took a long time for upper management to admit it and flush the disaster as soon as a better alternative appeared. Lotus Notes was one of those disasters, and it took a long time for it to die the death it deserved. Many users had to bear the heavy burden of its confusing UI and unpredictable behavior. Microsoft Word is another example of a user's nightmare (don't you just love the way it handles lists?), yet it persists, torturing each new generation of business users. And still, as late as the year 2013 A.D., as we saw with the costly and glitzy healthcare.gov website, there seems to be no end to the crap Tech is willing to put on the shoulders of users. (For a look back, see the Interface Hall of Shame[20].)

To all of these users I say, "I salute you, and I'm sorry for my part in it! It's not your fault."

And what of me? Initially, I bought into the whole *gestalt*. Yes, the UIs were confusing, but it was my job to explain them and, damn, I was going to be the best explainer in the world. There wasn't a UI that I couldn't figure out. If it took 23 steps across 6 screens to complete a task, I could figure that out, and I could document it clearly and concisely. If you could read, you could follow my steps and complete the task! Yes, we tech writers were the high priests and priestesses of the UI. We were the keepers of the UI secrets who deigned to reveal them in careful steps to the waiting users, who knelt worshipping at the foot of new Tech. We were dedicated to Tech and innovation, taking pride in coming up with new ways to reveal the secrets—Quick Reference Cards, Online Help, Training Videos, interactive CBTs, attractive graphic posters to illustrate the complexities of the screen sequences and the labyrinthine menu structures. Yes, users, when you looked upon the software in awe of its mystery, we would lift you up with our documentation. For within our words the secrets of the screens would be revealed unto you. Oh, we few, we band of brothers and sisters, we writers—how we wielded our power over them. No one else wanted to do it, that's for sure.

As someone who believes that Tech can make users' lives better (as it has mine), I couldn't help but notice that the car was heading into the ditch. It gradually became clear to me that much of the Tech we were documenting was not helping the people who were going to be using it. Whatever benefits the Tech had were not reaching its users. I finally had to admit that it would be much more effective to make an easy-to-use UI in the first place, instead of writing books that described how to use a confusing UI.

Let's be honest, no one wants to get new software and immediately sit down with a couple of books to figure out how to use the damn Tech. It's true that, in the era of the Lords, everyone had to "read the documentation" to use the software. But nobody liked doing that. "Why can't we just sit down and start using the damned thing?" those silly, naïve users wondered. Well, you could, if the new Tech just cared enough about you to employ user-focused UI design and coding. But it didn't, and the award-winning documentation work I had been doing was, for the most part, a waste of time. It just added to the burden on the user's shoulder.

All the effort, all the years of my purported success, and my real contribution was to provide very strong ropes for users so they could keep pulling, pulling, pulling that heavy wagon of unusable software down the road. "Look, my users, I've made you even more successful beasts of burden for Tech! You must learn to give the code what it wants—this is the way of the user."

And the worst part was that no one noticed their labors, or if they did, they didn't care.

Well, it was 2001 when I decided that I would not be a part of that ever again. That attitude was unacceptable. From now on, in my world, I had a new way of thinking, and I announced it plainly to anyone who would listen:

> ***The code is there only and always to serve the user's needs.***

Cool. But where was I to go with this new world view? I couldn't keep riding along with the Lords of Tech. On the other hand, I wasn't ready to get off the racetrack completely. I hated my tech writing job, but I still loved Tech. I still wanted to make it work for people. But how? Who would hire me with this unorthodox view about users? I felt caught in a net of despair.

And then, everything changed. Now isn't that just like Tech?

CH 5. www.whatjusthappened?.com

In the early 1990s, while on a tech writing contract at SAS Institute, a developer friend invited me and a few other folks to come to his office to see something interesting. He invoked a window on his workstation screen that called itself "MOSAIC." He explained that this window was being created dynamically on his screen through communication with a server located in a geographically distant place. The window contained text that explained what MOSAIC was, and some of the text was highlighted in blue and underlined. My friend told us the blue text was called a "hyperlink," then clicked it with his mouse, and poof! The original text on the window was immediately replaced with new text, which explained in detail what the hyperlink had referenced. Pretty cool.

Someone asked, "How do you make those MOSAIC pages and that hyperlink that you just clicked?" "Well, you use a tagging language called HTML," my friend replied. Hmmm. That piqued my interest. I had experience with tagging languages, and I had made hyperlinks in that old IBM BookManager program I worked with in 1991. I asked my friend to show me a sample of the HTML link that produced the page that we were looking at. He did, and OMG, the heavens parted, and the angels sang. I thought, "I'm already an expert!" HTML was obviously just a minor tweak of Script/BookMaster. Those hyperlinks were almost the same as those old BookManager links I had made!

Time passed. Over the next 6 years internet browsers came and went. We all learned about the power of linked online information. People with home computers and dialup services could now get more than just email and ListServs over the internet. They could browse the World Wide Web. By the late 90s, the number of web users had dramatically increased. People were browsing the web at home for their own purposes and at work for business purposes.

While at work, business users interacted with two kinds of websites—those that were outside their company, called "internet websites," and those that were inside their companies, called "intranet websites." By 1999, browsing either the internet or your company's intranet was often how you got the information you needed to complete a work task. Everyone in a company—from maintenance to management—was a web user, and they frequently were using the same web pages to get things done. As a result, there was no place for bad UIs to hide.

When links were broken, when they took you to the wrong page, when they took you to outdated information, they were quickly discovered and became well-known. When UIs on a web page were so confusing that people couldn't easily find the information they needed, complaints were loud and frequent. Users would get mad, and they made noise. They couldn't get their work done! Since management was often among those angry users, the noise was heard throughout the company. Suddenly, users mattered.

Companies typically established a group of developers dedicated to developing, maintaining, and updating the company intranet. But things were different for these Lords. They had to make the intranet work *for the users who were all around them.* It had to work right because if it didn't, their company's business efficiency and productivity would be affected adversely. If employees had problems with the intranet, the problems had to be fixed, period. And there were always problems because the Lords were not trained in developing UIs that satisfied users. They were trained to focus on the customer, and now, in a vicious turn of the tables, the customer and the user had become the same!

Developing web pages seemed like fun at first, but the Lords found that their intranets were constantly under fire for usability issues. Furthermore, the attitude that users were important was spreading to other development projects in the company. The day of the Lords was fast approaching an end

External internet websites also took fire for usability issues, with serious consequences. There was competition out there, and if your site lost users, you lost money. A perfect example of this was the fierce competition among web search engines. In the early days, the primary search engines were *Mosaic*, then *Netscape*, *Alta Vista*, and *Yahoo*, as well as minor players that cropped up from to time. They were all competing for users because they could make money through advertising on their search page. The more users they attracted to their search page, the more advertising they could win. Each company went into a frenzy trying to attract users to its search site. And who in the company was in charge of making the site attractive? Why, the S&M people, of course.

S&M departments were given serious funding to develop these search engine websites, and they went wild. All the search engine websites became jam packed with the latest HTML marketing ideas. Marquees, colorful graphics of cowboys and mountains and compasses, gifs, talking gifs, icons, talking icons, flashing icons, black pages with white text, blue pages with black text, two columns, three columns, no columns, top sections, middle sections, columns and sections with different colors, and advertisers' text boxes with hyperlinks scattered all over—-there was no end to the razzle-dazzle and jumble of HTML visual gimmicks that the S&M people put on these pages. Of course somewhere, somehow, a little bit of space had to be carved out on the page for the user to enter text and start a search. After all, that's why users come to the page in the first place, right? Well, not necessarily.

S&M seemed to think that, with the right combination of web page razzmatazz, they could capture users' attention and entice them to return for more, independent of a desire to search. This thinking lead to the addition of even more stuff on the search page. In particular, users started to see rectangles on the page filled with unrelated topics of interest, such as weather forecasts, stock market data, and breaking news headlines. These topics offered sparkly links, buttons, and sliders to tempt users to interact with them. Now the search page was even more dense, packed with both advertising and condensed information that might be of daily interest to users but had nothing to do with searching. With this ever-changing web page collage, it became difficult for users to find the search entry field on the page, if they actually needed to do a search. Here in this place today, moved somewhere else tomorrow. Soon the usability issues around searching were significant enough that business users complained.

At one point during one of my tech writing contracts, Yahoo had become the reluctant favorite. We used to say it was best to have your morning coffee *after* you did your Yahoo searching. The site was so slow to load and annoying to look at that if you mixed searching with caffeine, something on your desk was going to have a bad day. Staplers, squeaky drawers, small potted plants—they were all potential targets to relieve frustration. User needs were not being met by these search sites, and users were complaining. In this competitive age of web Tech, it was no surprise that someone noticed and figured out what to do about it.

I recall the morning I came to work to find the whole office buzzing about something. I kept hearing an unfamiliar word that sounded very silly. Probably just some developer's pranky joke, I thought. When I reached my desk and flipped on my workstation, a colleague ran up and excitedly asked me if I had tried the new search engine yet. I hadn't, and I asked what the big deal was. She said, "Just try it—-you'll see. Type in 'www.google.com'." I did, and when the site opened, I was presented with an almost empty white page containing just one display item on it!

In the top area of the page I saw a big, obvious, clean display of a text entry field followed by a Search button. There was nothing else on the page to distract me from the entry field and the Search button. They were sitting there patiently, waiting for me to use them, if I so desired. I felt relieved and calmed. All the white space surrounding the search field gave me confidence that nothing was going to interfere with or interrupt my search, if I chose to perform one. The site was giving me a focused, pleasant, visual invitation to do a search, without any distractions. So, I searched, and the return of results was also a great experience, with no distracting display of irrelevant information.

But there was more going on in the experience of this site. As I looked once again at the initial search page, I felt that the Google UI was sending me a powerful, subliminal message that said, "Hey there, User, what you want is important to us. We know you want to search the web, so we tried to make that task as easy and fast for you as we could. Here is all you need, nothing less and nothing more. Enjoy!"

The rest is history. Google's competitors rapidly dissolved into thin air. Finally, users started to matter to the bottom line of software companies. And if you go to google.com today, you will see the same, beautiful, simple, and friendly search UI. I guess we users still matter to Google.

The Web rapidly developed into a new way of doing business. Websites were free and open to anyone with an internet connection. Instead of ignoring users like most of Tech was doing, the owners of these sites worked hard to attract users and make them happy. Even though they often didn't understand how to get and keep users, they definitely kept trying—sometimes learning from their mistakes and making improvements. The time they spent on understanding users was worthwhile. After all, users could be the source of big money!

As for me, I looked up one last time from the brink of my disillusionment with the Lords and saw Tech make this great turnaround towards users. It was inspirational and energizing. Web Tech was now the new car roaring down the raceway. Running hot on HTML, eye candy, and JavaScript, this ride was leaving the Lords of Tech in the dust. I wasn't quite sure who was driving this hot rod, but I jumped aboard anyway.

5.1 That crazy thing called HTML

One crazy thing about HTML was that it was easy. Anyone could learn enough HTML to make a web page, and that made it fun. The early HTML tools required an author to use HTML tags to create a page. But simple web pages that contained basic text only needed a few tags, so it was not difficult to learn what tags to use to create the page. With web pages so fashionable and easy to make, many companies decided to engage employee loyalty by inviting employees to create their own biographical pages for the company intranet. Most employees participated, finding it easy to create a simple page with HTML tags.

Very soon, though, a variety of WYSIWYG web page authoring tools became available, so no one ever had to work with HTML tags to make a page. And these new tools also made it easy to insert the latest HTML and JavaScript eye candy on a page. With the availability of all this razzle dazzle, employees started to love playing with their company web pages and competing with colleagues for having the coolest page. Surfing the intranet and reading fellow employees web pages, then updating your own page, became an important part of the workday. Company intranets started to balloon with web pages full of every kind of HTML nonsense. Things quickly got out of hand.

People loved making web pages but had little interest in maintaining them. As a result, a company's intranet information was often out of date. Sometimes links were miscoded, either going to the wrong place, or to an out-of-date place, or else to no place at all. Upcoming page changes or additional information were sometimes promised but never fulfilled. The developers in charge of maintaining the company intranet decided to focus only on the mechanics of keeping it running. They wanted no part of verifying, editing, or designing either the content or the display of the information that was delivered over the intranet. Without maintenance of the information, chaos soon reigned over the company intranet.

The internet also started suffering from the same maintenance issues. The internet was informational, and every medium to large company had an internet website. The purpose of these company internet sites was to market the company, give an overview of what the company did, and provide details about the company's services. Competing companies added more detail and eye candy to attract potential customers away from their competitors. Competition lead to an increase in the number of interconnected pages on a website and deeper menu systems for navigating all those pages.

Of course, the more complex the website, the greater the maintenance requirements. Those in charge of the website, typically the company's S&M people, did not want to spend time updating and editing web content. As a result, it was not unusual for users to find dead links on internet web pages and confusing interconnections among a site's pages. "Where am I and how do I get back to the page about this product's details?" was a common complaint about internet websites.

Meanwhile, as the novelty of the web wore off, company employees began to lose interest in their personal intranet web pages. It was too much trouble to keep updating them, so eventually, they just ignored what they had created. The fad had run its course. At the same time, people began to lose faith in the value of their company intranet. Managers constantly heard their people complaining, "Our intranet is useless! You can't find anything that you need. You have to wade through pages of irrelevant or out of date stuff, and if you do find what you need and click the link, it doesn't work anymore or it takes you to worthless information!" Trying to send employees to the intranet for help or questions about their jobs or the company was not working.

In 1999, soon after I was hired to manage a 10-person documentation group, I realized that the company intranet was a confusing mess. I discussed the situation with the VP of Development and volunteered to fix the intranet by overseeing its content and design. I told him that I had expertise in HTML and information design, and I was interested in gaining experience in web design. I also saw it as a great opportunity to make some Tech that worked right for users—my new mission. The VP happily agreed. He needed someone to fix the mess.

With the help of a few interested senior writers in my group, I planned to edit and restructure the current intranet content as needed and become the gatekeeper for new content—in addition to managing my normal workload, of course. It was going to be a challenge, but I also had a lucky break. Through some quirk of interdepartmental politics, my group had inherited the S&M group's graphic designer. I gleefully involved her in providing a new graphical look and feel for my design of the new intranet, which she found to be an interesting opportunity and spent significant time on it after hours.

As we reviewed all of the current content, we found that much of it was employee and departmental bio pages that reeked of neglect. We informed all page owners that they were responsible for giving us updates to their pages, which we would edit and then post. And we stated that any broken links we found had to be fixed in a timely fashion or they would be removed. And, we added each page owner's name to the bottom of his or her page, so any visitor would know whom to contact for questions about the information on the page. I redesigned the main page with the basic company content and the menu system that led to the rest of the site. My team kept the main page and its links to company-based information updated.

This maintenance effort was time-consuming. In those days we had no automated web tools to test links and run spell checkers. Our efforts worked, though. The guys in charge of the backend running of the intranet did what we told them to: publish this stuff, delete that stuff, put these graphics in this folder. The site's navigation was clear, logical, and easy to use. The content was current. People could find what they were looking for. They never experienced the frustration of a broken link. They were attracted to our artist's graphic eye candy, too. These graphics were professional looking with an *avant garde* flavor that got people talking, even if they weren't always fans of the style. Our company intranet became useful again, and I had become the Tech Lord of the intranet—at least until the company was bought out, and we lost our jobs.

One lesson I learned from my experience of being a Tech Lord of that relatively small intranet was that it was a full-time job. Sure, HTML was easy, but doing websites right took various talents, significant time, and perpetual maintenance. And I wasn't the only one to realize this. An entirely new Tech business quickly came along to provide website design, content management, and ongoing site maintenance for companies who didn't want to do it themselves. These web services were available for either internet, intranet, or both. And one of the benefits these companies promised was—wait for it—*usability*. "Yes, folks, with our web services, your website users will have easy navigation and direct access to all of your important information. Your web pages will always be up to date and easy to read, using the latest features of HTML and JavaScript. And the graphics we use will be professionally designed and rendered to delight the eyes of any visitor to your site."

Web design became the new Tech, and usability was a big part of its vision.

5.2 The usability profession exits the race

Once again, companies doled out money for a new Tech vision. "Yes, yes," they said, "we need someone to fix our websites so users will visit them! We need usability!" They knew, of course, that they needed websites that were easy and delightful for their target audience to use. What these companies didn't know was how to get there. So, they bought into the vision: the Usability Professional knows how to get you there.

Usability Professionals had been around since the Golden Age of IBM. That company wanted its hardware and software products tested for quality and customer acceptability *before* they were sold or released to customers. IBM invested heavily in usability departments made up of operations and industrial engineers as well as Ph.D. psychologists. If you worked on an IBM mainframe project, you couldn't swing a dead mouse without hitting one of the Usability

Professionals on the project team. These folks understood hardware, software, and users, and they didn't hold back on their analyses of a product. At IBM, the Usability Professionals on the team often had final sign off on a product. It couldn't be released until it passed usability.

When the web's cry for usability hit the market, some big Tech companies decided to create their own internal usability departments, instead of outsourcing their website development. Microsoft still has world class people with stellar academic credentials in the field of product usability. So, how do these folks work to try and make software easy to use? Was the vision of usability the answer to creating great websites? What is the work of a Usability Professional, exactly? I'll digress for a few pages to answer these questions with a little history.

In the Tech world, usability analysis is focused on the behavior of a software application while someone is using it. The usability professional analyzes a software application using several methods. In one method, the analyst observes users working with the software and examines the degree to which these users find the software's behavior easy to understand and helpful for completing their desired tasks. The result of this analysis is some type of report that explains which features of the software are acceptable and which are problematic and should be fixed. If only a few minor features are problematic and are easy to fix, the software's usability would be rated as acceptable. As the number and severity of problematic features increases, the Usability Professional's rating of the software trends toward rejection. In other words, the software's usability is so poor that people should not use it.

Observation of users is one method for evaluating usability. Another method is heuristic analysis. With this method, the usability analyst uses an application himself or herself, exploring all of its features. During this exploration, the analyst measures how well the software conforms to a well-accepted set of usability heuristics (usability rules). For example, one important usability heuristic is that the navigation system of a software application should be logically organized and use clear terminology that is meaningful to the target users. To analyze an application's navigation system, a usability analyst would use the navigation system and evaluate how well it conforms to this heuristic. The result of the heuristic analysis is a report that explains which features of the software conform to usability heuristics, which features do not conform, and the degree of their non-conformance. The report may also define a usability conformance scale and provide a usability rating based on that scale.

The takeaway here is that a usability analysis can tell you what is right with a software application, what is wrong with it, and how serious its usability issues are ***only after*** the software is developed and ready to be used. That was a big problem for the Lords of Tech, and I have some sympathy for them. If the software they are ready to deliver is shown to have serious usability problems, will the Lords want to trash most of their work and start over? The S&M people had a similar problem with the discovery of usability issues near the end of the project. Will they be willing to go back to the customer and explain that users won't like the software, and it needs to be reworked? Of course not. That's not how they do things. Users are not their focus.

Before the rise of the web, I was amazed to see that there were lots of usability analysts working in software companies. And most of them did their jobs. If newly developed software had serious usability problems, they found those problems and reported them to the project manager. But here's the problem: it never mattered what the usability analysts said. The responses they received were "Thanks. We'll try to fix some of these problems in the next release," or "Thanks, but we can still release the software. The customer won't care about these problems. If necessary, we'll provide workarounds in the Readme file."

(Ah yes, the Readme file—the file of the damned. In it you unfortunate users will find the list of all the features that don't work right in the software and what you can or, more likely can't, do about them. These screwups are euphemistically referred to as "known bugs." Yes, users, you've just paid for something that doesn't work right, but we're giving it to you anyway. Enough! Let's don't get me started on the Readme file[21].)

And so, until the web came along, the Usability Professionals did their jobs and, aside from the Golden Age of IBM, they were largely ignored. Despite their efforts, it was clear that they weren't going to be able to make software significantly better for users. Software that violated any number of usability rules continued to be released. (No surprise, for example, that we still suffer with Word.) But when the web design companies saw that they could sell usability, whoosh! the usability Phoenix rose from the ashes.

Two things happened. First, it became fashionable for web users to point out the usability problems with websites and applications. Everyone experienced problems on the web, and many found fame reporting them to the world on their blogs (this was pre-YouTube days). Usability Professionals were especially vocal, since they were a big part of the new wave and finally seemed to be getting their props. And they were right to expose the errors. A favorite usability reporting site of mine was The Interface Hall of Shame, which is still somewhat preserved in various places on the web.

The second thing that happened was that the new web design companies quickly discovered the limits of usability analysis. For the most part, it simply wasn't going to work for them. The new and exciting Web Design Tech car ran on race gas. Websites had to be designed and deployed rapidly, then updated frequently with the latest information and, of course, the eye candy currently in fashion. That was part of the vision these companies were selling. There simply wasn't time to design, implement, deploy, do usability analysis with target users, then redesign, reimplement, and re-analyze to fix any usability problems that were uncovered. Web design companies needed to find a faster way to build a high degree of usability into their sites without waiting until the software was finished. The way of the Usability Professional was not the answer.

In my own quest for user satisfaction, I initially started down the Usability Professional path, but soon realized that usability analysis could not lead to great software. More was needed than just saying what was wrong with an existing software application's behavior. I needed to be able to describe, in detail, what the software should be doing to satisfy its target users ***before the coding started***. My problem was that I wasn't sure how to do that. Fortunately, with the web's new focus on user satisfaction, some amazing ideas about how to make user-friendly software were hitting the Tech waves. I found these ideas exhilarating, and they led me down an innovative and fruitful path.

5.3 UI Design: breaking new ground

In 1998 my husband handed me a recently published book about designing user interfaces that he thought I might like. It was written by a developer! It changed my thinking and my career trajectory. That book was *About Face*[22], by Alan Cooper (now in its fourth edition). The book sketched out a way to make great software for users without relying on usability analysis. Simply put, Cooper's idea was this: the tasks that users need to do with a software application and the ways they need to do those tasks must be learned and designed into new software from the start.

With that idea as my foundation, the way forward started to take shape. If user satisfaction was bound up with user interface design, then there were two tasks I had to complete: first, I had to devise a process for learning what target users needed from a proposed, new application. And after I completed that task, I needed to formulate a reliable method for designing those user needs into the appearance and behavior of the new application's screens.

These two tasks challenged and engaged me at a high level. To devise a method for capturing user needs, I drew heavily from my technical writing experience. I had written many step-by-step user scenarios for completing tasks. Now I needed to figure out what kinds of questions I should ask users if I actually had the chance to talk to them before I wrote their scenarios. When it came to designing screens, I practiced for hours with blank sheets of paper, a pencil, a ruler, and some sample user scenarios I wrote up. With many months of practice, I came to understand which parts of the UI needed to be designed first and why. Once I had those important lessons down cold, the rest of my UI design process fell into place.

Soon I was at the point where I could storyboard an application's screens on paper. That was extremely satisfying, but I knew it would not be enough to make a difference for user satisfaction. For my design efforts to work, I needed to be able to convince a project team to let me design the UI *before* the developers coded it. And if they agreed to that, I then needed to be ready to present the team with more than a basic storyboard of boxes and circles on white paper. I needed to show them a design that would engage them and look real enough to prove it could be coded.

To accomplish that, I first tried using various graphics programs or PowerPoint slides to make realistic screen designs based on my crude paper designs. I created pictures that mimicked the appearance of real screens with color, UI objects, and readable screen text. But it was still difficult to demonstrate to others how these static screens would behave and flow from one another. I worried over this problem for awhile, and then the solution hit me: I can use that crazy HTML!

Managing the company intranet had taught me a lot about how to code screen actions with HTML and a bit of JavaScript. And I had recently learned how to use DreamWeaver, which made it easy to create HTML pages with active JavaScript objects, such as buttons and menus. That was all I needed to make an interactive set of web pages that behaved like real application screens even though nothing was really happening underneath them. With that crazy HTML I could rig my web pages as interlinked screens that behaved the way users needed to complete their tasks. I didn't have to make every function or feature on the screens do something, but I had to put enough interactivity into the design to show the flow of the most important user tasks through the sample screens.

What I made is now called a "high-fidelity, interactive UI prototype." I just called it a prototype in those early days. For those projects where the development group agreed to let me design the UI, my prototypes were very persuasive to the teams, especially to the stakeholders. These prototypes showed that users could accomplish what they needed to do with an easy and satisfying sequence of actions. All along, I had known it was the UI design that stakeholders needed to see to evaluate the software, not the dreaded pages upon pages of use cases. The stakeholders and user testers would pore over my interactive prototype, asking questions and offering additions or modifications. Here was something they could relate to and work with.

Sometimes the prototype was so realistic that the stakeholders and user testers fell into thinking it was the actual application. "Why doesn't that work?" they would ask, pointing to a menu or button on the screen. I would have to remind them that the screens were only a prototype and not everything worked. "And that's a good thing," I would say, "because we can easily make design changes now since the coding hasn't started yet!" Then I would explain to them what I thought the UI object they had asked about should do, and get their feedback. We would discuss what changes, if any, they thought should be made. By the end of the prototype review, they would be sold. "Yes," they would say, "that's what we want!"

My designs were never perfect, and my methods for gathering user information were always improving, but I had made the journey to the light. Now I knew what had to be done to make software that satisfied and delighted users, and I knew how to do it. By late 2001, I was breaking new ground as a very early entry into a burgeoning new career area of Tech. What was I? How did I bill myself on my resume? Who was willing to hire me? All good questions, mostly for the next chapter. For now, suffice it to say that I had found the way forward for achieving my goal: user satisfaction. Looking back, it's clear to me that I couldn't have done it without that crazy thing called HTML.

5.4 Help! We need some user satisfaction!

Despite my forward motion with successful UI design methodology, Web design companies were still struggling with how to build good usability into their designs. They had accepted the new ideas from Cooper and others about building user satisfaction into the design process. Their problem, however, was finding people who knew how to do it. They turned first to the Usability Professionals, "Can you help us put usability into our website design process?" they asked. The usability folks replied, "You mean you want us to start with a blank sheet of paper and draw UI screens that we know the users need and will like?" "Yes, yes, that's what we want," said the web design companies. The usability folks turned away, saying, "Oh, sorry. We don't do that. We evaluate software after it's designed and working. Bye now." (BTW: This response was the beginning of the end of the usability profession. If you search for it now, the Usability Professionals Association (UPA)[23] has since rebranded and restructured itself as a user experience design group.)

Outside of developers who coded UIs, there were no professional UI designers that the web companies could find and hire. The discipline didn't yet exist. Now in a world of hurt, the web design companies next turned to their in-house professional graphic designers. After all, they were the ones who made all of the eye candy. Surely, they must know what users like. "Hey, can you guys design the UI so it will have good usability?" The answer should have been "No," but some graphics professionals thought they could figure this out using color and cool graphical objects. It didn't take long for the web design companies to see that the graphics pros had no clue about user needs and ideal user task flows. The websites they designed were pretty but still as confusing and frustrating for users as ever. Now what?

Next up were the S&M people. “Hey you guys understand how to sell to users, can you design the screens so they have good usability?” The S&M folks were all over it. Their designs became one big marketing mashup, like the old days of Yahoo. Clearly, that wasn’t going to work. It was time for the web design companies to think outside the box.

How about these new business analysts we’ve been hiring? Yes, Business Analyst (or BA for short) was a new career track that had emerged within the Cleanup Crew. The S&M people had decided to abandon the burden of requirements gathering, and the BA position developed as an offshoot of S&M to take on that task. As it evolved, the BA became responsible for documenting requirements and ensuring that both customer and project team were in agreement.

Typically, BAs understood little about the Tech their projects were all about, but they were personable and accommodating. They documented what stakeholders told them, passed documents around, then updated them to the project document repository as needed until sign off was achieved. Of course, they never talked to users or even asked to talk to users. They accepted whatever management told them about users and wrote it down. Most often, they worked under the direction of the project manager.

When the web design company asked them to design screens, they were personable and accommodating, but had no clue what to do. “What do they want us to do?” they asked their project managers. The most they could get as an answer was “Just draw some rectangles with buttons and menus inside.” They tried their best, but their rectangles were always rejected by the customer and project team. “We need training!” They told their project managers. But alas! There were no schools or accredited courses around that trained people in user-focused UI design.

The last stop was the Project Manager, but this was never a realistic possibility. The PMs were too busy and announced clearly and loudly that they had neither the skills nor the interest to do UI design. There was no one left to ask except the technical writers, who actually understood the concepts of user needs and gathering information from users. Yet the writers were passed over as if they weren't there. Perhaps this was because the web design companies didn't create much book-like documentation. They made websites, which contained all the text they needed online.

And so, the developers in the web design companies continued to control the UI, with almost the same disastrous usability results as the Lords of Tech had churned out for workstation applications. Fortunately, two new pieces of Tech had come into play to improve things for website users.

The first new Tech was a set of unofficial but mutually agreed upon UI rules that every website developer followed, for example, the behavior of the Back and Forward buttons, and the mouse buttons. There was also agreement about checking for and eliminating dead links on a website, and developers made tools (web spiders[24]) to automate this rule. Conformity to these basic rules not only helped the coders but also the users. Even if the rules didn't always facilitate task flow for a user, at least the user could rely on the consistent behavior of at least some website UI controls and behaviors.

The second new Tech was the development of templates. The cowboy time of web development was coming to an end. Web design companies had done their research. It became clear that some website colors and layouts received more user views and interactions than others. Soon, every company wanted to have one of the "good" designs for its website, and the website design companies were able to crank these out from the basic templates that they created.

The templates became a great way for the web design companies to make fast and easy money. With templates in place, they soon found that really great developers seeking elegant code (like the Lords of Tech) weren't needed to develop websites. With templates, there wasn't much to it, so they no longer hired the best and brightest. Finally, even though the usability was still not great, at least users were able to learn a specific style of website, which made it easier to know how to manage the usability problems they encountered. One of the more famous parts of these website templates was the Site Map. This map was essentially a web page containing an index of all of the pages in the website. So, when users got lost, they learned to go directly to the Site Map and search for the page they were trying to find. It wasn't fun, but it was better than nothing.

It was great to see things improve for users, but so many problems remained. At least there was some awareness of the fact that pleasing users could directly improve a company's success. This awareness spread dramatically when a new player showed up on the Web at the start of the new millennium. E-commerce arrived and stole the show.

5.5 From Search and Browse to Buy and Sell

Ebay, PayPal, Amazon: these companies hit the scene with websites offering a whole new way of thinking about shopping. They brought us *online transactions*—buying and selling—and took the web to a new level. The initial search-and-browse-information model of the web could not support transactions. Now, money could change hands and goods could be transferred through computer interactions with a website. A user could sit at the computer in old, ratty pajamas and buy new ones without leaving the house. Incredible!

From the coding and web design perspective, online transactions were not easy to enable and support. They demanded a whole new set of Tech. Security, database account management, order tracking, delivery logistics, inventory control, returns, customer loyalty—the list of Tech's new web coding needs and backend software architecture todos seemed endless. And things had to be done right, or big legal issues and maybe even bankruptcy ensued. When money changes hands, the whole game gets as serious as a heart attack. The days of just throwing up a template were not going to cut it for the transactional websites.

Fortunately, the Lords were still around, and many of them were up to the task. Online companies arose that did an amazing job of handling transactions over the web. They typically started small and built up their Tech as they grew, learning how to do it better with each iteration. And all of it had to be done in double time. Rapid innovation and deployment schedules were still the norm for websites. The backend coding for a transactional website was where the innovation often took place. Database development, money movement, and order and delivery logistics all had to be drawn together in a tight web of security. Mistakes were made, and they were always costly. The usability of transactional sites was typically very poor, but if a user could get through a site's purchasing screens and its credit card security worked, it was a life-changing experience for consumers.

Soon almost every retail company with a physical store was also trying to sell online, and many of them failed. Competition was fierce, and sound website Tech was critical. More than ever, the need to please users with the online shopping experience came to the forefront. With so many companies selling online, consumers could price shop more efficiently and with little effort. If price wasn't a factor for the consumer's choice of seller, then shipping costs, return policies, order communication, delivery reliability and timeliness all came into play. Only a few companies floated to the top, and they got most of the business because they did the best for users. Amazon was one of the companies that users came to prefer.

As Amazon grew beyond its book selling business, it decided to perfect the user's buying interaction. Even today there is no better model of a satisfying and delightful online purchasing experience than buying something from Amazon. And this includes the whole package—from secure and easy online purchase through delivery and return. Other companies have tried to match it but few, if any, have succeeded. They all fall down on one or two of the many things that Amazon does so well. Why?

When I analyze the Amazon experience, I see the best of the Lords of Tech at work, creating their elegant code in rapid fashion. And I see excellent UI designers perfecting the user interactions on the screens. (I infer from this that these designers have successfully wrested control of the UI's design from the Lords.) Other companies have either not made that kind of investment in UI designers and Tech Lords, or have not set up their web project teams and processes to favor the satisfaction of their target users.

Amazon aside, the trend towards templates for transactional websites moved quickly. While many of these templates still have usability issues, they do incorporate standards for backend security, database accuracy for account management, and logistics and tracking modules. Some web design companies now provide templates for both transactional websites and informational websites. Fortunately, a degree of stability exists for consumers who buy and sell over the web, but it took years and many online transaction failures to get where we are today.

The worst usability offenders now are government websites, from local sites all the way up to federal. These groups don't face competition and aren't accountable to their users. But they aren't the only ones that give users bad headaches. What kinds of usability issues do we typically find on poorly designed sites? Here are a few of the most egregious issues, which I'm confident have victimized all of my dear readers at one time or another.

Common Website Usability Issues

- I can't delete an item in my cart!

- I just want to buy one item and never come back, so why are you making me fill in 3 pages of forms to create an account?

- I had to come up with a password to create an account on your site. You gave me no rules for the password. After I put one in, you showed me a big, red error like I was supposed to know what your password rules are! You trap me into making a mistake by not giving me the rules, then you act like I'm stupid and finally tell me the rules like you're doing me a favor!

- I spent 10 minutes filling in your multi-page account form and *after* I clicked Submit, you decided to tell me I did something wrong. Of course, you didn't tell me what I did and where I made the mistake!

- I was filling out a multi-page online form. Part way through, I realized I had made one mistake on a previous page. I completed the page I was on, then went back to the previous page and fixed my mistake. When I went forward to the page where I had been, ALL my input was deleted, and I had to start the whole page over again!

- After I submitted my order, you charged an unexpected fee, and your Customer Support told me there was an option I failed to check on the ordering page. I looked back at that page, and eventually saw the option and checkbox at the bottom of the page in tiny print and pale blue color that no one could see!

- You won't tell me the shipping fee until AFTER I submit my order!
- You sent me a coupon for your site but I can't figure out how to use it.
- I finally completed my order and clicked Submit, then your site timed out with no information or status. I don't know if my order went through or not!
- I was on page 3 of your form when I realized I needed to change something on page 1, but you only let me go back to page 2!
- I can't find the Return Policy on your ordering page.
- I checked the option to refuse your marketing emails but I keep getting sales reminders and your Unsubscribe link doesn't work!
- Fifteen minutes after I completed my online order, I realized it was wrong, but you wouldn't let me cancel it!
- The delivery date promised on your ordering page was off by 3 weeks!

Fortunately, users continue to fight back with a powerful tool that the web itself has made possible: online reviews. Never before have so many people been able to warn so many other people, across the world, about bad customer experiences and pass on favorable recommendations for good customer experiences. While there are abuses and trickery in the online review process, it is possible to filter most of that out and separate the wheat from the chaff. Users helping users is becoming the gold standard for identifying websites with usability problems. User reviews may not be the only factor affecting the success of a business, but they have become a major factor in evaluating the usability of its website. And a website that is repeatedly reviewed as hard to use will definitely have fewer visits from potential customers. Thanks to user reviews, businesses often attempt to fix their usability problems. Unfortunately, governments still don't seem to care and continue to force old Tech and sloppy Tech on their captive audiences.

As the web became more and more "templatized," most e-commerce websites eventually looked much the same. To be sure, there was coding safety in proven e-commerce templates, which now included the backend transaction code. And templates also saved companies money by avoiding any custom coding or innovation that went beyond interactive eye candy graphics. With the arrival of templates for e-commerce, the more brilliant Lords of Tech were no longer in demand for website development. But another opportunity arose for them.

A lot of money had been made very quickly by small e-commerce business owners. Now, they were ready to make even more by investing in new Tech ideas for online services and products that they could sell online. These investors cast a wide net with their money, and many Lords of Tech blew into their offices to sell them on exciting, but typically only half-baked, product ideas. No problem. That was just the kind of opportunity the investors were looking for. Not too much money required upfront, a lion's share of the potential profits, quick development of a prototype, and then on to glory! The Lords took the investors' money, and the brief, cutthroat, and incredibly fruitful era of Tech startups began.

5.6 Bubblemania: startup creativity vs. cookie-cutter code

The success of the web changed the way companies thought about software for their business. The web browser was so easy to use, and it was the conduit to online customers. An online e-commerce website became a whole new way to increase the customer base of a business.

Companies like banks, who wanted to start transacting business on the web, started either by buying online banking applications or developing them in-house. Since their websites were public sites, every bank could see every other bank's online application. This capability led not only to fierce competition but also to imitation. The banking execs would tell the software company's S&M people that they wanted an online banking web application, and furthermore, that they wanted it to do everything their competitors' applications did, even to the point of looking like their competitors, only better. This demand for web-based business applications that all followed the same model of appearance and interaction drove Tech companies toward standardization and templatization. And because they were web applications, banks were demanding faster turnaround times.

Remember the Lords of Tech, seeking innovation and elegant code with their huge development projects that could stretch across years? Well, those days were gone. And after doing the hard work to develop solid, secure backend e-commerce websites, they were no longer needed. The new e-commerce templates took the Lords' innovative work and standardized it. Big business customers, like banks, could buy out of the box, cookie cutter code that worked like everyone else's, was fast to create, fast to update, and fast to modify as new functionality came along. So, with the new millennium barely in the rear-view mirror, if a Lord of Tech wanted to do something interesting, innovative, challenging, and cutting edge, which would probably require elegant code, he had to find a startup opportunity. And there were so many to choose from!

Amazing Tech, developed with elegant code and sparkling with innovation, began to flow out of startups. Some of it never saw the light of day because the investors' money ran out. But a lot of it hit the market with big fireworks. (Here are some examples[25].) Every week saw some new Tech product announcement from an obscure startup that no one had ever heard of. "IPO" became an everyday term. Would this one become the next big thing? Maybe. People were making lots of money buying into new startups, getting in on the ground floor. At the peak of this frenzy, even big players like IBM and Cisco decided to participate. Millions of investment dollars were flowing from big Tech companies to new startups of unproven value. Take a chance! Even if just one of them pans out, we'll make all our money back, and the product will be ours!

The templatization of web-based business applications and the startup frenzy created a need for UI designers, yet finding the right person to fulfill that role was an ongoing problem. No one understood the skills and methodologies required for user-focused UI design. As a result, a variety of professionals with ill-suited skills were hired for the UI design role, and they mostly paid attention to eye candy, information mapping, focus groups, and user surveys.

Users were still suffering because UI design was not understood in so many ways. Not only was the nature of a UI Design role unclear, but there was confusion about how a UI Designer and the design process fit into the project team and the software development lifecycle. Who should have final control of the UI design? How much coding knowledge should a UI Designer have? Will the UI Designer slow down the project schedule?

These were important issues and valid questions. I worked on my user-focused UI design methodology and identified the tools I needed. I also defined the skills required for user-focused UI design, and I developed a presentation of this information together with a convincing rationale for hiring a UI Designer. I could answer the questions a hiring company might ask about the role, and I could lay out how the role worked with the project team and the successes the role could bring to the project.

I had completed my journey to the light. I was armed and ready to work with the Lords so that together, we could bring Tech to users in a way that would satisfy and delight them. But just as I completed the final polish of my UI Designer resume, a loud POP stopped all Tech cars on the raceway. Tech's enormous, floating bubble of startups and big company web application development had burst. Suddenly, the shining glass and marble buildings devoted to Tech software companies were surrounded by empty parking lots. The landscapers were nowhere to be seen. Instead of manicured flower gardens, For Lease signs were planted forlornly in front of building entrances. Tech's grandiose office parks became ghost towns.

Big Tech companies and small Tech startups were scrambling to avoid bankruptcy. These companies had either been overvalued or valued for factors that were irrelevant. As the true value of the Tech market started to reveal itself, the hordes of hungry Tech investors began to slip away like sand through the hands. In the blink of an eye, the Lords of Tech had skittered and scattered in the wind, while the Cleanup Crews were laid off, with little hope of finding new Tech gigs. It was a bad scene. It was called The Tech Bubble.

One of the more interesting lessons I learned from the few startups I worked in was that you don't necessarily need 15 developers and a 2-year project cycle to come up with amazing, life-changing, innovative Tech. Just one, or maybe a few brilliant developers with an ingenious idea, when left to themselves—working out of a closet-sized room or someone's home garage—can produce incredible Tech in only few months. Their code will need some polishing and their screens will need some good UI design, but they sure didn't need a passel of S&M people, layers of management, and a 15-month timeline for a project team saddled with useless requirements, low level coders, and assorted hangers-on. This lesson did not go unnoticed by Tech companies, leading to a wider interest in and acceptance of new ideas for how Tech projects should be run (for example, small, co-located project teams, Extreme Programming, and other agile software development methodologies).

Even though the Tech Bubble of 2001 left a lot of roadkill on the raceway, no one doubted that the world would continue to demand the software innovation that startups delivered. Despite the Bubble's carnage, the gold rush of Tech startups continues even now. The risk is daunting, but the sweet smell of success remains irresistible. Like everyone else who was laid off during the Bubble, I knew it was just a hiccup, and Tech would recover shortly. And I knew that user-focused UI design was going to be needed to make great software. Unfortunately, I seemed to be the only one around who knew it!

Despite the recent, universal realization that users mattered, Tech was not ready for me. To get the jobs I wanted and do the important work I was now ready to do, I would have to win the hearts and minds of developers and their managers. Could I convince them to let me control some of their sacred turf?

CH 6. Can't Get No User Satisfaction?

It was time for me to leave pit lane, get on the raceway, and see if I could attract some attention. I was still a Tech Consultant in the Cleanup Crew—my work still fed off those who developed software. Without them there wouldn't be any users for me to satisfy! Now, though, I was going to break some of the rules governing Cleanup Crews. First off, I would be in my own Tech car, and I would be driving. Second, I would carry passengers, but not the usual crowd. The passengers in my car would be target users. Whenever I worked on a software project, I would bring along the target users for that software. And third, these users wouldn't be there just to say Hey y'all. I would make these target users integral to the project team from start to finish.

When it was time to tell the project team what the software needs do, the users in my car would be first in line to provide the requirements. They would be there during development, too, to verify that the software was meeting their requirements, and they would be there at the end to give the final Go or No Go. My vehicle was a *Users First* Tech Car, and no one on the raceway had ever seen one before. But anyone I passed would see that my car was all about user-focused UI Design, even if they didn't know what the heck that was. For sure, that car turned out to be a doozy. And I've been driving it since 2002, upgrading as I go.

I outfitted the car with my own company, called *Focused Design.* It was a simple Sole Proprietorship, but that was all I needed. I designed and implemented my company website, designed and printed some business cards, then revved the engine and threw her into gear. Filled with hope, we zipped down the entrance ramp and merged confidently into the jammin' traffic. To my delight, I snagged a good number of job interviews. But it took a good year for me to figure out what needed to be said during those interviews to get the job on my terms. The companies I interviewed with knew they needed help to improve user satisfaction, but they didn't know why it was important to their bottom line, how it should be done, or who needed to do it. In other words, they didn't know what my role needed to be. So, I had to tell them. I had to explain how I could build user satisfaction into their software. Some of them didn't like what I had to say or didn't understand it, but as time went on, I got better at explaining what I do and why they needed what I do. It wasn't long before most of my interviews resulted in a job offer. "Hey Hey Hey That's What I Say"

6.1 Job Interview Part 1: the WHY of user satisfaction

Once I realized that my interviewers may not even know why they needed me, I restructured my interview strategy to tackle that issue. At some point in the interview, the interviewers began asking me open-ended questions, such as "Given our description of the position and its responsibilities, how do you think you can contribute to our projects?" I would respond with something like "My value is the focus I can put on ensuring that the software you make will satisfy and, moreover, delight your users. Remember, ease of use of software and satisfaction with software comes from the way the user interacts with the software screens. In other words, user satisfaction comes from the right UI design of the screens."

After this type of statement, I would continue with something like "Your description of the position emphasizes the goal of user satisfaction, which looks like a good match to my interest and skills. Can you talk a bit about the level of importance this goal has in your organization and among your project teams, and how you've been working towards that goal so far?" Their answer to this would tell me if they had a clue what user satisfaction means, who among their company's management actually cares about it, and their idea of how to achieve it. Based on their response, I would launch into as much explanation as needed about why user satisfaction is a critical-path goal in a software project, and why the organization must provide strong support for that goal. I would usually spend time discussing how user satisfaction affects their business success and bottom line, because I knew that was a huge concern for them. When I was finished, I would assess whether they were with me. If they weren't, we were all wasting our time, and I knew I would not accept the position if they later offered it. If they liked what I had to say—and they could meet my price—the chances were good I'd be joining their team on contract.

Obviously, my explanation of the "why" was a key step in the interview. I had to be sure my explanation was to the point, logical, complete, concise, and easy to follow. I was not trying to persuade them to see things my way. Instead, my goal was to explain clearly and honestly what I do and how it could benefit them. I had to probe deep enough to know whether they understood my role and, if so, would support me. I also wanted them to be able to make an accurate assessment of how well I matched what they wanted. It took a while to get there, but I eventually was able to structure my "why" explanation so that I could learn what I needed from the interviewers and get the job, if it looked like a match. Here are the points I typically covered in an interview, paying close attention to the interviewers' reactions and comments.

User satisfaction inspires customer loyalty

A loyal customer base is like gold to any company. If your customers are loyal, they come to you first for products, services, and advice. They come to you before even considering your competitors because they know they can count on you. An important added benefit is the good word they put out about you. Your loyal customers continue to do business with you, *and* they often bring you new business. They are highly prized, and that is why companies will devote time and resources to finding and keeping loyal customers.

In the business software Tech market today, research has demonstrated that the traditional methods used to gain loyal customers are no longer sufficient. For example, sales account relationships, favorable pricing structures, onsite training and support, extensive online help and documentation, call center support, rapid response to problems with workarounds and follow-up releases—these are all now easily undercut if your customer experiences reduced employee productivity because your software is hard to use. If the target users of your software can't get their jobs done right or in the expected timeframe, your customers will start to look for a better solution.

When your customers start looking at your competitors, that's the first step toward losing them forever. If they decide to buy somewhere else and that new company solves their user productivity problem, we know that it will be much, much harder to gain their trust again. Look what happened to Yahoo when Google came along! Look what happened to Barnes & Noble when Amazon came along! And, remember the death of Lotus Notes after Microsoft Outlook came out? Finally, don't forget the collapse of American cars and the rise of Japanese cars! The losers in these battles for customers failed to provide the user satisfaction that the winners gave us.

You may have already learned this, I would continue. You may have seen that expensive investments in free training, detailed documentation, and call-center support for your software, plus quick problem solving with "workarounds," isn't helping much with keeping your customers loyal. Research has shown why these things don't satisfy users when they are having problems with their software. One reason is that users don't like to spend time in training. They know it's hard to remember everything from a few training sessions. And they know that the training will not cover the trickier situations they will inevitably face when trying to accomplish their specific tasks with the new software. Another reason is that users are easily frustrated if they frequently have to refer to a book when they are having a problem trying to complete a task with the software. Finally, today's user expects software to be easy to use *the first time,* especially if you are providing web-based software. You can't ask your customers to wait for the next release. Their users expect it to work right *the first time.* If the new software doesn't satisfy users right away, management will start to look elsewhere for software that will.

In today's market, a customer may like you, but if that customer's employees don't like your software, your friendship will not be enough to hold that customer. And these days there is increased downside to losing a customer because the word of mouth about your software will go viral. A few online complaints or some negative online reviews can reach millions of potential customers in minutes. Even worse, what's on the internet has a way of staying on the internet. As a result, the hill you have to climb to win back a customer is now a death-defying mountain. That's why user satisfaction is critical to your bottom line, and the market forecast says that the effect of user satisfaction on customer loyalty is expected to increase over time.

As I always tell my prospective employers, the path to user satisfaction requires a commitment to *getting it right the first time.* If our software projects don't commit to getting it right the first time, we will fall victim to all of the excuses that lead to hard-to-use software. As for understanding what users need the first time they use an application, I know how to gain this understanding and put it into the software. That is how I can help you achieve your goal of user satisfaction. That is how I can contribute to your project.

ROI from user satisfaction

At this point, I stop to take a breath. If they are not with me, they already have or will now want to argue about or deny the points I've just made. In that case, I don't waste too much more time with the interview and make a graceful exit (you never know—they may come around later). But if they are with me, they will take the next step and ask me how I work. "How do you plan to gain the understanding of users that you say is needed, and how do you imbed that understanding into the software to make it work right the first time?" In response I tell them that I have developed a method and process that I know works. But before getting into the specifics, I ask if I can briefly address the issue of ROI (return on investment) in user satisfaction. They perk right up, because ROI is always in the back of their minds, hanging over their heads like the Grim Reaper.

Metrics, metrics, metrics—software company managers love metrics that quantify how much a process, policy, or methodology contributes to the company's bottom line. For example, suppose a development manager goes to his VP and asks him to make a big investment in Agile software development methods. The development manager talks about how much more productive the team will be if they can start using these new methods. "You want to start using an Agile development process instead of our current process?" the VP says, "Well I need metrics that show how much this investment will increase company revenue, in terms of time and resources."

Of course, coming up with such metrics is only a shot in the dark, since many of the most significant benefits of process change are not quantifiable. With this in mind, I started my discussion of ROI by boldly stating that metrics are not the way to measure the ROI in user satisfaction. For example, we know that increasing user satisfaction helps increase customer loyalty, and we know that customer loyalty can improve company sales. But by how much exactly does user satisfaction help revenue? How quantifiable is that claim? Marketeers often turn to customer surveys to find out why a customer has purchased a product. But such surveys do not provide metrics. If done correctly, they can provide true indicators of customer thinking, and such indicators can help confirm or refute the success of a product or idea. But metrics you can put in your budget are not available from surveys. The value of strong user satisfaction cannot be denied, but neither can it be quantified.

However, the lack of hard and fast metrics for measuring the ROI in user satisfaction doesn't mean it can't be measured at all. One important way to measure it is by contrasting the effects of investing in improving user satisfaction with the effects of releasing products with poor user satisfaction. Here is an example I would often use in interviews to illustrate this method of measuring ROI.

Product Parameters Affecting Revenue	With Strong User Satisfaction	With Poor User Satisfaction
customer loyalty	Better	Worse
support burden and customer complaints	Less	More
demands for software fixes and "workarounds" post-release	Fewer	More
demands for user aids (training, documentation, online help, quick reference materials, readmes)	Decrease	Increase
ability to meet future product release dates	Better	Worse
online reviews and word of mouth	Better	Worse

When I discussed the information in this table with interviewers, they could not disagree with it. The truth of it is self-evident. Which software product would a company like to make—the one with strong user satisfaction, or the one with poor user satisfaction?

This table also shows, in a shocking fashion, that metrics are not required to measure the ROI of user satisfaction. It doesn't matter "exactly how much" user satisfaction improves these product parameters. It's abundantly clear that improving them to any degree is going to be more valuable to the company's bottom line than settling for poor user satisfaction and its associated costs. In this new day of Tech, everyone had accepted that users mattered. Companies could no longer afford to ignore them. In interviews, this table was my go-to illustration of why hard and fast metrics are not needed to measure the ROI in user satisfaction, and in fact, they are not even appropriate.

So much for the why. If the interviewers were still with me, it was now time to explain how I worked to guarantee user satisfaction with software.

6.2 Job Interview Part 2: the HOW of user satisfaction

I began the discussion of this point by stating the obvious. To achieve user satisfaction, we need the right software design. The UI has to provide ease of use. But to know what the right design is, we must first understand our target users. Of course, since they are people, there are so many things we could learn about each of them. More things, certainly, than we would ever have time to find out, and more things than we would ever need to create the right UI design. So then, before we begin trying to understand users, we must know what we need to know about them. We must be focused on getting the right information and not wasting time on irrelevancies.

"Here," I would say to the interviewers, "are the questions we need answered to get what we need to know about our target business users." I would then present the following graphical representation of these questions, based on the standard software project where the goal is to replace the user's current software with something better. (If the goal is not replacement but instead providing a software solution for the first time, I would discuss the minor differences in focus.)

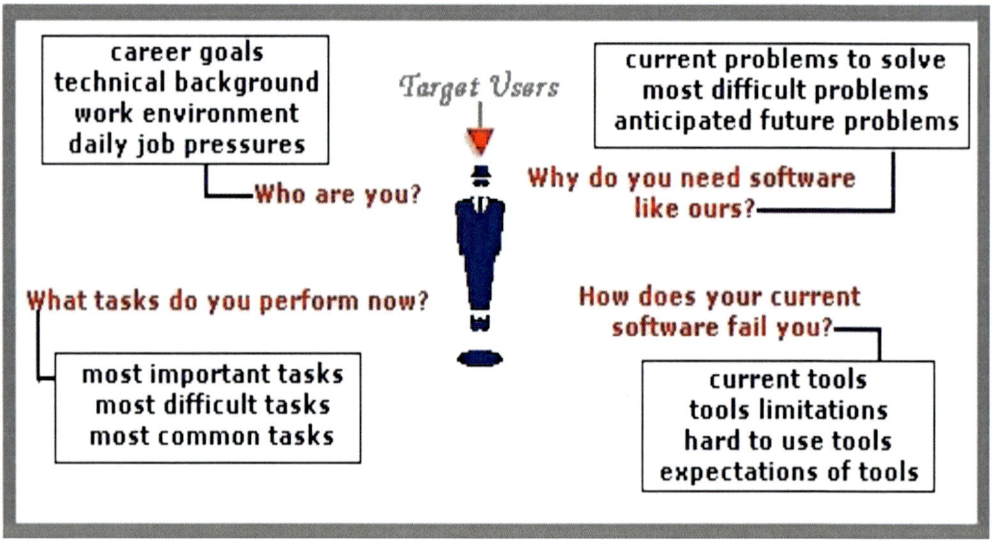

As I worked through the questions represented in this graphic, I would explain how they help us understand what users need and why. But I didn't have to spend much time on this part, because all present could readily envision how these questions would get us to the heart of what users needed in their software. To seal the deal, I would ask them to do their own thought experiment: "Suppose your company was going to replace the current software you use for reporting your work hours on various projects. If someone asked you these questions about how you use the current software, do you think the questioning would uncover what was needed to increase your satisfaction?"

After they acknowledged the power of these questions, I addressed the next issue that I knew was on their minds: How do you go about getting users to answer these questions? I also knew the standard options they thought I was going to suggest, but they were in for a shock. Those options are very far off the mark from what's really needed, and I needed to explain why. So, I ran through the standard options quickly, pointing out their limitations and weaknesses. Then, after I had taken away everything that they thought could work, they were a bit more ready to accept the shock of what was needed to get the right information from target users.

Standard Option—A. User Surveys

1. Format Limitations of User Surveys:
    - Restricted to a short list of questions because users are unlikely to complete a long list.
    - Each question must be short and unequivocally phrased, avoiding more than one sentence or multiple ideas.
    - Questions are restricted to one-word answers, such as Yes/No/Maybe, Sometimes/Never/Frequently, Agree/Disagree

2. Resulting Failures of User Surveys:
    - The format does not allow immediate, follow-up discussion with a relevant human being of any points or ideas that users have related to the question.
    - The format does not allow for users who are interested in the idea of the question but don't like the way the question is formulated.
    - Most comments are not useful, even if the survey has a comments section. Users don't want to elaborate on their answers in a survey because they don't want to take the time to formulate a precise statement about their ideas, feelings, and recommendations. It's too much effort for no matching reward.
    - Most users believe that completing the survey will not result in the improvements they really need with the software.

- The questions are specific rather than open ended, which leads to a bias from the questioners towards information *they* think is important, which may or may not match what users think is important.
- The brevity of a survey limits the amount and the complexity of the information that can be gathered.
- The actual sampling of users is difficult to control, which makes the results unreliable. Even if the survey is distributed evenly across the groups within the sample, not everyone will complete the survey. The actual sampling can turn out to be too small for generalizations about the users. Further, the number of people in each group who complete the survey will vary. That means the surveys that are returned will not carry the same statistical weighting across groups. Figuring out what the survey actually shows for the given sample is, therefore, problematic. Any inferences based on the survey results are likely to be weak, at best. Almost certainly, they will not be strong enough to show us what the right UI design should look like.

Standard Option—B. User Focus Groups

1. Format Limitations of User Focus Groups:
   - Because there are multiple users in the group, each user is limited in how much he or she can speak on a particular subject.
   - Users cannot demonstrate what they are describing about the software because they are not in their daily work environments.
   - An issue from one user may be appear to be less important if other users in the group have not experienced it.
   - The comfort level with speaking in groups differs with each user.

2. Resulting Failures of User Focus Groups:
   - Group dynamics can skew the results since the loudest, most emotional, most assertive, and/or most frequent voices will be heard most. The introverts in the group will be intimidated and reluctant to speak against them, or else become so frustrated with the process that they just check out. While the moderator can attempt to control the louder voices, they will always carry more weight since they are heard more than the others.
   - Disagreement among users as to the relevance or importance of one user's viewpoint results in that viewpoint carrying less weight, even though it may in fact be a significant viewpoint that the others, simply

by chance of their random selection, don't understand or haven't had the opportunity to experience in their daily work. The risk is not small that something critical can be missed.

- Emotional responses to the moderator's general questions about preferences for individual UI components (for example, "What are your thoughts about using slider bars?") are not reliable measures of the value of a UI component in the context of performing a particular task.
- Individual users who need a long time to express an important point are likely to be cut off before they are finished.
- Users who find it too complicated to express an important point in words do not have the opportunity to show what they mean by actually using the software as they would in the context of their daily work environment. As a result, the important point will be lost.
- What people are willing to reveal in a one-on-one conversation is typically much more useful than what they are willing to offer in a group discussion. As a result, critical user information that we need to create the right UI design will be missed in focus groups.
- Focus groups work best when the question for the group is simple and unrelated to their work or personal performance. For example, a perfume company may want to test individual scents in a focus group. "Which scent do you like best?" is a simple and nonintrusive question. On the other hand, complex questions about software operation that are specific to how individual people do their work can be intimidating in a group setting. People don't want their work issues to be judged by their peers or strangers.

Standard Option—C. Usability Testing

1. Format Limitations of Usability Testing:
    - Usability testing is conducted in a laboratory environment to allow hidden recording and observation of user behavior. This environment differs dramatically from the user's daily work environment.
    - The usability tests are required to test the most common, most difficult, and most important scenarios that the users will face with the software. Therefore, detailed knowledge of these scenarios must be gathered before the testing is done.
    - Because of laboratory costs and user availability, testing is time limited.

- The software used for testing must be ready for deployment to ensure valid testing of scenarios.
- Usability analysis describes the problems that users have with the software but does not provide UI designs that will solve those problems.

2. Resulting Failures of Usability Testing:
   - Lack of familiarity and test anxiety in the laboratory setting skew the user's performance. As a result, the conclusions drawn from the testing lack reliability.
   - Time limitations preclude testing complex user scenarios, which typically involve many steps. But not testing these complex scenarios misses gathering critical information.
   - Because the software is already completed, the results of the tests are highly unlikely to affect any major software changes before deployment, even if the results show that such changes are needed.
   - Knowledge of the relevant user scenarios that should be tested is required before testing. And that knowledge only comes from asking the kinds of questions I already listed. Since usability testing comes after the software is designed, gathering the relevant knowledge about users at that point will not be used when it needs to be used: during the UI design process.

Standard Option—D. Letting Users Design the UI

I have participated in JAD sessions where developers asked the business representatives in the group (or specific users they invited to the JAD session) to suggest screen designs that would meet their needs. Since these folks had no experience or expertise in UI design, nothing useful resulted. Moreover, these questions often frustrated and even angered them, and I fully sympathized with their reactions. Designing the UI is not their role. They can't do it, and they don't want to. Their contributions are to tell the team what tasks they need to perform with the new software and to describe the problems they currently have performing those tasks. They can prioritize these tasks by importance and identify the most critical problems that need to be fixed. If they have knowledge gaps, they can give the team access to their business users who can fill in the gaps. These contributions are vital, and only the business representatives can provide them. Expecting them to design the software screens that would solve their problems is like expecting the owner of a house to design the blueprints for the new addition he wants built. Ridiculous.

Standard Option—E. Use Cases

I have already described use cases and discussed their limitations in Trick 2 on page 59. Suffice it to say that use cases provide too much information that is irrelevant to the user's logic while failing to understand how users think about their daily tasks. In the end, creating use cases takes a great deal of time and provides little guidance for how the screens should appear and behave to satisfy users.

So much for the standard options. So, what does work? After my review of what won't work, I quickly move to the good news, and at this point, my audience is typically all ears. Now is the time in the interview when I have a chance to change the *status quo* and describe how to design software that satisfies and delights users. Now is the moment I've worked so hard for, and I was definitely ready. My passion would run high throughout the rest of the interview, sometimes too much, but I wasn't going to hide it, even if I could. What you see is what you get. That only seemed fair to a potential employer.

My Method—A. Structured, One-on-One User Meetings

I would begin my discussion with the obvious—software projects move fast, so we don't have time and money to spend on things that don't work. Fortunately, the method I've developed will get us the answers to the list of critical user questions and enable us to create the right UI design. As I explain this method, you will be able to see how my method avoids the failure points of the standard options we've already discussed. The first step to user satisfaction, as I said, is to gather the information we need from target users and do it right the first time.

To make users comfortable, I meet with each one in his or her daily work environment, whatever that may be—a cube, an office, or the sales floor. If the users are geographically distant, I will travel to them. Since they are comfortable and familiar with their surroundings, they can show me what they are talking about as well as tell me. To respect their time constraints, I ask for 45-60 minutes of their time, whenever it is convenient for their schedule, and I never exceed that time unless they ask me to stay.

My preference is to have only two of us meeting with a user—I manage the interaction, and the other person is the scribe who takes notes. Having a scribe accomplishes three critical goals:

- First, the scribe frees me up to move through the meeting without having to take time to write down important information.
- Second, since the scribe is focused only on the notes, the scribe will be able to take more complete notes than I could.
- Third, the scribe hears and sees everything I do, which enables the scribe to serve as an additional observer. This backup greatly improves the reliability of our conclusions when we later review what the user meant when making various statements and explanations. Simply recording a meeting and returning to the recording later, after having met with many users, is a much less reliable method of drawing conclusions about what a user meant at the time of the meeting.

To ensure that I make the most of my time during the user meetings, I prepare a worksheet of questions. The interviewers are often curious about this worksheet. I tell them that, essentially, the worksheet contains the questions on the generic list that we discussed earlier, although I typically make them specific to the domain and software we are focusing on. My scribe becomes familiar with the worksheet and uses it as a basis for capturing information during the user meeting. Well before the user meetings, I research the relevant domain and review the software the users already have. I also review whatever information we already have about our users from all available sources (such as stakeholders, S&M documents, domain research, and reports from the customer support group, if there is one). When I do meet with users, I am already immersed in their context so that I will be able to understand the basics of what they do.

To start a user meeting, I thank the user for his or her time and explain that the project team I am working with wants to understand what is needed to solve the problems he or she has with the current software. I tell the user that this information will help us create the right software design. Rest assured, I say, that I will not use your name or identify you in any way when we analyze the information gathered from this meeting and the meetings we are having with other users. I will leave my contact information with you so you can feel free to send me any additional thoughts or ideas or any questions. Do you have any questions at this point? After answering any questions that the user has, I proceed.

I begin with my first question from the worksheet. From there, things always get both interesting and enlightening. No matter how many user meetings I've completed, and how much I've already learned about my target users, it's always an awesome confirmation of my method when I gain a brand new, important insight during the next meeting. We can never learn everything, but we must learn enough to identify the primary goals and needs of our users. If we can satisfy just those and no more, we will be extraordinarily successful.

My meetings with users require me to listen more than speak, and to request elaboration (can you show me?) when needed. I have to know when to dive deeper and when to move on or change direction. I have to be sensitive to what the user wants to talk about and indulge him or her as much as possible, exploring whether there is important relevance hidden under the surface. I call this method of interacting with users *Guided Listening*. The basics of the method can be learned, but like anything, being really good at it requires specialized insight and talent. When done well, Guided Listening always results in getting the information needed.

During my job interviews, I would come prepared to hold a sample Guided Listening mini-session with one of the interviewers so all of them could observe the method's effectiveness. Once in a while, they would try it. Whether or not we did that, they all became believers. But they had concerns.

How many users do you need to talk to? How much time does this take to complete? How much does it cost? We can't pay for you to travel around the country or around the globe to talk to our users!!!

All valid concerns, which I was prepared to address, for the most part. I would tell the interviewers that I need to identify 10-20 users who are representative of the most important types of users your new software must satisfy. Specifically, I need a cross-section of user skill levels, company levels, and functional roles. For example, I might need basic users and advanced users; sales associates and managers; data entry users and data analysts. Once identified, we ask for an equal number of users in each type. This doesn't need to be exact, but the closer we get to at least 4-5 users of each type, the more reliable our results will be.

The interviewers are now taking notes on the details of what I'm saying. I continue by describing my experience with user meetings, which is that after talking to 4-5 users of each type, most of the information I learn starts to become repetitive. This is an important signal that my information gathering task is approaching completion. Once the level of repetition is high, I know I have enough of what I need to move to the next step of the method.

As for time and related costs, my experience is that it usually takes about a week to identify the types of users, find real users of those types who agree to meet with me, and then schedule the meetings to occur over the next few weeks. I typically complete all user meetings in two weeks. It saves time and costs if the users I need are all in one location. However, if geography is a factor that we must consider in identifying user types, then travel will almost always be required. In rare cases, I say, I've been able to conduct a user meeting through a web meeting service, but this is not suitable for most users.

I can see skepticism in the eyes of the interviewers. Perhaps they are thinking that all this user contact is overkill. Do we really need all that? I emphasize that the danger of using my method is in cutting corners, for example, thinking that we only need to talk to a few users, or that we can meet over the phone or video chat with users instead of in person, or that we don't need a scribe to be involved. Cutting corners is dangerous because it causes degrees of separation from our users; it creates walls between us and our users, thus compromising the amount and the quality of the information we get. We've already discussed what happens when you don't get the right information. So, I say to the interviewers, we must be faithful to my method, and if we are, I can guarantee the results.

At this point, I would pause for a moment, because I knew this was a hard pill for them to swallow. No matter how you sliced it, giving me the time and money to talk to users at the start of a project was going to be something they had to defend to management. OTOH, the reason we were all here was to find something and someone who could solve a problem that management wanted solved. I gave them a few moments of silence, and after taking a few deep breaths, they came back to me. OK, they would say, suppose you get that access to users and get the necessary information from them, what's next? I smiled. Now, I could tell them how I go from Guided Listening in user meetings to a UI design that users love.

My Method—B. Synthesis and Analysis of User Data

The next step in my method dealt with a problem common to any analysis: now that I've gathered a mountain of data about 10-20 users, how do I sort through it all and come up with valid inferences about what users need and what will satisfy them? How do I decide what user data is critical and what isn't? How should I structure the data to facilitate reliable analysis?

This was an important problem for me to discuss in my interviews, because I couldn't go to a project team with a set of software screens I had designed and say, "Well, I've talked to a lot of users, and I think they will want the UI to look like this." That kind of approach will never fly with developers. They will frown and say, "Why? What's your evidence that users want it like this? I've talked to a few users, and I don't think they would want it like that!" That's what they'll say, and I wouldn't blame them. They were the ones who had to write code that worked, and they deserved better from the UI Designer.

I had worked hard to develop an organized, repeatable, and fruitful method for analyzing the data I gathered through user meetings. I told my interviewers that my method demonstrated that the conclusions reached by my analysis were based on verifiable user data and rational inferences from that data. As a result, members of the project team, especially developers, would have no reason to challenge my analysis as subjective opinion or guesses. The team could have confidence that my analysis used a clearly described procedure that was grounded in reliable data.

Similarly, if someone on the team asked me why the screens should look a particular way, I could point directly back to the user data that supported my design. I emphasized to the interviewers how important this would be for building trust with the team as I presented screen designs. They all agreed that how I analyzed the user data I collected would be key to being accepted by the team as a UI designer, and they asked for details on my method of analysis. I delved into the details, but only to the level they wanted, given the time available in the interview. I gave them documents that went deeper into the finer points of the method, should they want to read them later.

I began by explaining that my analysis method is built upon a report template that I created to record data after a user meeting. As part of my method, I complete a user report within 24-36 hours after a user meeting, while the experience is still fresh in my mind. I base the report on the detailed notes taken during the meeting and the discussion I have about the meeting with my scribe.

The table below shows the sections in my user report template. This template organizes the user data into sections that reflect the general worksheet questions I use at the meetings. This commonality ensures that the information from the user meeting will be reliably preserved and transferred to the report.

My User Report Template	
Section 1	Primary user goals related to the user's job and daily work
Section 2	Detailed description of the user's job responsibilities and daily work environment
Section 3	Step by step description of the user's daily tasks related to the software of interest in this order: most important tasks most frequent tasks most difficult tasks easiest tasks including the user's reasons for importance, difficulty, or ease of use
Section 4	Most important current problems the user has with the software of interest
Section 5	Key needs of the user that will improve daily work and help achieve goals, based on current problems and the kind of changes seen as helpful. For example, a user would identify a key need by saying "It's so easy to get lost in the screens because there are so many steps in this critical task that I need to do every day. If I could somehow see where I am in the whole process, it would really help me keep track of what I've completed and how much more I need to do."

Another significant advantage of using my report template is that, after completing the user reports, all of the data is organized into the same structure. Some of the sections use a narrative format while others require a detailed list. But when I look through the reports, I can see the same kind of information arranged in the same format, rather than having each report written in a different style and/or organization.

With these well-structured user reports in hand, I can begin the analysis. I start by creating a spreadsheet that has a column for each section of the user report and a row for each user. To fill in the spreadsheet with information, I read through each report one section at a time. I distill the information from each section, then put it into the appropriate column of the spreadsheet in the row for that user. When all user data has been added to the spreadsheet, I can review the data down each column, looking for similarities and differences. The revelations that are uncovered in this review are remarkably informative.

For example, my spreadsheet review identifies the software needs held in common among the users I met with. These common needs are the ones that need to be analyzed. How important are they? To answer this critical question, I look to the column that contains the key user needs. "Which of these key needs are also held in common?" I ask myself. Those are the needs that are most important, and they will become the major requirements that our new software must meet to achieve the greatest user satisfaction.

The point is that my spreadsheet review and analysis allow me to identify the user requirements for our new software and prioritize them. I am able to put these requirements into the following categories:

- **Key needs**: the requirements that must be met for the software to satisfy most of the target users

- **Relevant needs**: the requirements that, if met, would be highly appreciated by most users

- **Unilateral needs**: the requirements that we shouldn't try to meet because most users don't need them

I tell the interviewers that once I've finished my analysis, it's time for me to present our target users and their requirements to the project team. My opening statement is, "I want to introduce you to the people our new software needs to satisfy and tell you what they said they need." At this project meeting, I synthesize the users in the spreadsheet down to one or two personalities. I describe them to the team, including their goals, daily tasks, and key needs. I present the prioritized list of requirements, emphasizing that users never expect to get everything they want in new software, but they do hope to get the most important things they need.

Delving into the requirements, I tell the team that if we can at least implement the major requirements on the list, we will satisfy our target users. And if, in addition, we can add a few of the other requirements that they would find helpful, we will delight them. What does that mean? It means they will say to their management, "I love this new software. It solves the problems that were driving me crazy, and it knows what features are really helpful. This software truly understands what I need to get my job done faster and better!!"

Now the interviewers see how the parts of my method fit together—the user meetings, followed by user reports, and the spreadsheet analysis based on those reports. They can see that, throughout my process, the connection to the user remains unbroken. There is no way for unnecessary requirements or irrelevant feature requests to enter the analysis. The conclusions that emerge from my analysis about what users need are cemented in what I have learned directly from users.

The interviewers can also see how I focus the project team on the user and provide the team with a reliable analysis of user data. Now the team has enough confidence in the requirements to move forward with specific software features and UI designs. When we contrast my method with the old ways of gathering software requirements, it's beyond obvious which method is more likely to result in a successful project that achieves user satisfaction.

By now the interview time we have left is getting short, but I still have to complete the "how" part of the interview. I move through it quickly and leave detailed documentation for the group to review later. Sometimes they call me to return another day to discuss the finer points, particularly with one or more of their developers. In any case, it's time in this interview to complete the story of how I get from the analysis of requirements to specific UI designs that meet those requirements. It's all part of my method.

My Method—C. From User Scenarios to UI Design

Getting the right set of requirements is the first focus of my method. However, even with those are in hand, we still need a design procedure that we know will meet the requirements in a way that satisfies our target users. For example, suppose our requirements show that users need a way to convert their online reports to PDF format. That may define the feature we need to code into the software, but it doesn't tell us how the feature should be presented to users so it is easy to use and available when they need it during a task. Suppose we hide the access to the feature inside a menu named "Other". Users will quickly forget where the feature is and have to spend time looking for it again and again. So, the rhetorical question I put to the interviewers is, "How does my method ensure that the UI designs I create will be the ones that users will like?"

To answer this question, I remind them that my method focuses on user tasks. First the reports, and then the spreadsheet show the steps users take currently to accomplish their most important and most common tasks. The steps for each major task comprise what I call the "user's scenario" for that task. Since I know that users must still perform those same tasks with our new software, my design process begins by looking at each step in a current user scenario. I work on understanding how that step can be improved by the new features we want to put into the software. Essentially, I work to create new user scenarios that include the new features that are relevant to the steps in that scenario's task. Each of my new user scenarios will state, in words, what steps the user will take to complete the task at hand, including how the user will work with the new features relevant to that task.

This part of my process ensures that I embed the new software features into the flow of the tasks that they relate to. Any step that uses a new feature must follow logically and directly from the previous step. Any new step must also lead logically and directly to the next step. And the logic I refer to is the user's logic—how the user is thinking in terms of his or her goals for completing that task.

For example, if we are going to add a PDF conversion feature for the online reports the user creates, I will create a new scenario for the task of creating an online report. I will begin by examining each step in the user's current scenario for creating an online report. I'll decide whether that step should be eliminated, added to the new scenario as is, or be modified to improve the new flow that will contain the new feature we are providing. As I move through the steps of the new flow, I'll decide where in the flow the user will want to access the PDF conversion feature, and then create the step at that point. The new step will describe in detail how the user will want to access the feature—through a menu system or a button or a tab, where the access point should be relative to what the user is currently viewing, and what the name of the feature will be, based on the user's viewpoint of the feature. There will be no screen representations at this point—only verbal descriptions of the scenarios.

Once I complete the new user scenarios, they serve as a functional description of the software features for the project team. The scenarios show the sequence in which the user will want to access the features and the points within a task that the user will need them. From a UI design perspective, this information tells me the visual level at which the features need to be available on the screen. It also tells me what kind of UI object the user needs to access the feature, such as a menu system or button. The new user scenarios essentially tell me in words what the target users want. With them in hand, I have what I need to create the UI design.

In the design phase I sit down with blank sheets of paper and draw the screens that will enable the user scenarios and only those scenarios. I do not add screen elements unless they are needed to enable the steps in the scenarios. But I don't just start with any user scenario. Instead, my method uses a top-down design approach. That means that I begin with what the user will see first when they use the new software. My first screens account for how the user will start the new software and the visual appearance of the startup, followed by a design of the main screen of the software. This is what the user sees first, and as we know, the first impression is critical. The main screen must show the functionality that the user needs after startup, and everything on that screen must be self-explanatory. The main screen must clearly display the access points to the initial tasks the user wants to do. This access could be a menu system or some other framework.

I call these initial, top-down screen designs the "Conceptual Framework", and the rest of the screens flow logically from it. Because the Conceptual Framework design is the foundation for the entire UI design, I present it to the team for feedback. I am prepared to explain why each element of the design needs to look as it does, referring back to the scenarios and, if needed, the user reports. My explanation is always based on what users have stated they need—never on my ideas of what they would like. I present all the details of how the screens work and will enable the user scenario for startup.

The project team will always have comments and suggestions, and whenever feasible, I accept them. I know that there can be more than one design that will achieve the same objective, so I'm willing to listen to other ideas. At the same time, I'm the expert in the room, and I'm the one who understands the target users and advocates for them. So, as a rule, I don't change my designs unless I'm convinced that another one is better. The team usually realizes this early on during my presentations and accepts it without incident. Developers typically focus more on identifying any gotchas with coding my new screen designs, and this meeting gives them a chance to bring those forward so we can discuss possible remedies. While another iteration may be required to nail down my Conceptual Framework, I've found that the project team is usually anxious to move forward with coding it. No one has to worry about dreaming up some UI at the last minute. And, they are relieved that the UI designer has shown herself capable of providing accurate screen representations based on a deep and thorough understanding of user needs.

I tell the interviewers that the remainder of the design work is a matter of adding screens that support the flow of the user scenarios. With an accurate and detailed Conceptual Framework to work from, this design effort moves quickly, including minor rework of any completed screens as needed. As the screens for a scenario are completed, they are presented to the team for review and coding.

I conclude the discussion of my top down design approach by contrasting it with the typical method of UI design that I've experienced, which is a bottom up approach. The bottom up method, I explain, assigns chunks of coding work to each developer, based on the features the software is supposed to contain. Each developer works on his chunk, with little concern for how his chunk will eventually be integrated into the whole program. Integration comes later. With this method, there is little to no emphasis on how the chunks should fit together in the main UI screens, since those are done last. As I point out to my interview group, the bottom up approach requires making the parts before you know what the whole should look like. With this approach, the chance is high that you will deliver software that is hard to use. Building a UI from the bottom up is like building the rooms of a new house before the architect creates the design for the whole house. That's a disaster waiting to happen.

Now, I tell the interviewers, there is only one more step to describe. That step is the failsafe of my method. Before the coding of my UI designs progresses too far, it's important for the team to know that the UI is on the right track. No one wants any surprises when the coding is close to completion. I need to be sure the designs are right, while there's still time to make major changes. This step is what guarantees that my method will create a UI design that's right the first time.

My Method—D. Design verification

When the screens are designed for a few of the most important user scenarios, it's time to create a high-fidelity interactive prototype and bring it to users for testing. This step must be done before coding of the UI is too far along, in case users have major problems with the UI in the prototype. I call this final step of my method "Design Verification". My prototype is just smoke and mirrors, but it works well enough that users typically think it's the real thing. I like to have 5-8 verification sessions, scheduling some of the users I had meetings with as well as some users I've never met. If my original sample of users was sound, the comments from all of the users testing the prototype will be consistent, with no outliers. These comments will uncover any common problems as well as common favorable reactions.

I structure the test much like a usability test. The test questions are the steps from the user scenarios. Essentially, I'm asking the users to try to perform their most important tasks with the new UI. The results will tell me if I've captured the user scenarios correctly, and if my UI design makes it easy for users to perform those scenarios. Either I or a helper sit with each user during the test and explore any issues at the time they occur. I also invite 1-2 stakeholders or members of the project team to observe and listen from an adjacent room, out of the user's view. The comments I receive are extremely valuable. I capture them in the form of a rating scale of 1 to 5 for each step that the test asks the user to perform. If a rating is not 4 or 5, I ask the users why and record their answers in detail. To conclude the user verification testing, I tabulate the test results and deliver a report to the project team.

The report tells the team what is right with the design, provides recommendations for simple modifications that will improve some things, and identifies any major problem areas early enough to fix the design before it is fully coded. In my experience, the testing results have been overwhelmingly positive. Most of the time, the test results reinforce my expectation that the Conceptual Framework is on track. Frequently, the comments are extremely favorable overall, even when users did not like a few aspects of the UI. For me, the typically favorable results from design verification confirm that my UI Design method does what I claim it does: namely, my method reliably results in user satisfaction. As for the project team, when they review the report and see the mostly positive reactions of the target users, they breathe a sigh of relief. Often, for the first time in a long time, they start to get excited about working on a software project. And they realize the value of the time I spent on meeting with users.

As I conclude the discussion of my method, my job interview is usually hitting a positive note. Everyone looks suitably impressed and is smiling like I'm probably the one. But I know they can't end the interview yet. Two serious questions always seem to be floating around inside one or the other of the interviewers' heads, and they always asked them, although somewhat delicately and indirectly. Straight up, the questions were these:

> *Do you have enough experience?*

and,

> *Your UI design method is all well and good, but how will you get them to let you do it?*

I felt I could handle the first question OK with a little dancing around. For the second one, though, all I could muster when I was just starting out as a UI Designer were a few S&M-type sound bytes, delivered with a lot of bravado.

6.3 Job Interview Part 3: the WHO of user satisfaction

It was 2001 when I first hit the raceway as a UI Designer. While the web had proved the value of good UI design, most companies, as I've said, didn't know what to look for in hiring a designer. Should the primary skills of the designer lay in graphic design, or information architecture, or business analysis, or marketing, or customer support, or development, or even some combination of those things? Companies had never had a UI Designer on staff before. In the past, all screen work had been managed by development. But now the world of company websites and browser-based software was focused heavily on how the screens looked, and development groups were not really into that. So, hiring some kind of designer to create the look and feel of the screens seemed to be the way to go, even though the required skill set for this work was somewhat mysterious.

My detailed explanation of my method helped convince job interviewers that I could design screens. However, they looked at my background and saw primarily technical writing, management, and a Ph.D. in philosophy. Whatever they were expecting for a UI Designer's skill set, that was not it. This was a problem for me, mostly when I first started looking for UI design work. I knew that I didn't have a strong job history in UI design, so I focused on providing the next best thing—peer recognition.

During the two years that I was preparing for my career change to UI design, I turned my ideas about design methodology and design rationale into published papers and conference presentations. After all, I knew I had developed valuable design ideas, and I had the academic skills to communicate them in a professional manner that others could understand and appreciate. So, as I worked through the development of my design method, I wrote about it as well, getting my work accepted into professional journal papers and conferences. UI design was becoming a fashionable topic in those venues, and my ideas were welcomed.

With a strong list of quality publications and presentations in UI design on my resume (as well as my company website, which I designed and implemented myself), I stood out in the crowd of competing job-seekers. In my interviews, I could claim valid expertise in the field, even though I was short on actual experience. In the beginning, most companies were reluctant to hire someone who had never held the job title of UI Designer or at least done UI design work on multiple software projects. But a few companies realized, after interviewing many candidates, that there simply weren't any people out there who had a lot of experience as a UI Designer. It was a new field of Tech work, and they would have to take a chance on someone. After discussing my peer recognition with them and distributing copies of some of my papers, I could tell that the interviewers decided I looked like a pretty good bet. (For interested readers, AP 4 on page 241 contains a redacted list of peer recognition items from my early days as a UI Designer.)

Once I had actual UI design work under my belt, I could provide a suitable work history and a portfolio of my UI design work. Although I continued to publish in the field, the peer recognition part of my resume became pretty much icing on the cake. When it came to my portfolio, the hiring companies mainly wanted to verify that I had a portfolio. But I wanted to do more than flip through a PowerPoint of screens or breeze through the pages of a website I had worked on. I always took the time to explain at least one of the samples in my portfolio to show them I wasn't cheating. I had learned from years in tech writing that the devil is in the details, and if you can't explain the details, you can't claim the work as your own.

When interviewers asked to see my portfolio, I would select a sample and explain the design goals I was working with for that sample, the key needs of the users, and why I chose the specific elements and arrangements on the screens. In this way, I demonstrated to the interviewers that the sample was indeed my work, and that I have the skills to create and present a design clearly and logically to others. (To this day it surprises me how many interviewers fail to demand this kind of demonstration of their design candidates' portfolios. It's a sure-fire way to learn if the candidate has done the work and truly has the requisite design skills.). Even though much of my design work dealt with proprietary systems, I was still allowed to show screens and website prototypes of my designs during my interviews. Of course, I couldn't circulate or leave copies of this work with the interviewers. But they understood this and respected my caution.

In any case, in the final part of my discussion about how to achieve user satisfaction though UI design, I would tell the interviewers that I have some questions for them. They always perked up at this because they liked to see that I was interested in how their company works, and it gave them a chance to describe it. But my questions were really designed to draw them out. I knew they had those two questions of their own lurking in the back of their minds, and I wanted to be sure they were answered.

Here is how that Q&A typically went in the last 20 minutes of the interview:

Q. Will I be given access to users for user meetings?
A. We would have to make that happen. It's a new idea for us, but we are confident that we can provide you the access you need, one way or another.

Q. Will I have access to your test system to provide tests for the UI, even if I have to do the testing myself?
A. We don't see why not. But we can talk to development and get back to you quickly on that.

Q. Will I have the ability to enter UI issues in your software bug tracking system and assign the level of severity to them?
A. Yes, we can talk to development to make that happen.

Q. Will the project team commit to regular presentations and reviews of the UI design as I progress through the design process for the screens?
A. Sure, the team is excited to have a UI Designer on board and very interested in working with you. If you are hired, we will schedule some initial orientation meetings with the whole team and also with specific individuals on the team for more in-depth conversations about your role. Proper integration into the team is important.

Q. Will I be reporting to a development manager or through a different department in the company?

With this question, sudden silence would quiet the room. Instead of a direct answer, one of the interviewers would usually hum and haw for a moment, then launch into what was clearly a sensitive and worrisome topic for the group. Now it was getting real.

6.4 Getting along with "them"

I knew what they were worried about. I had been working in the midst of it for more than ten years. The concern they were trying to communicate, as delicately as possible, was this:

> *Development controls the projects at our company, and the project team works within the boundaries that development sets. Whoever is your manager, no decision about your work is beyond development's governance. Developers are the company's most important resource, and everyone else on the team is expected to work successfully with them.* ***Can you follow your method and still get along with them?***

Their concern was valid, because they could see that my method encroached on responsibilities usually assigned to development. I was stepping on toes. I was treading on turf that belonged to *them*. I needed to put the interviewers at ease about this concern, but the ugly truth was that I did, in fact, want to take over control of development's UI design turf. I was prepared to step all over those development toes. Of course, I knew that a turf war with their prize employees was the last thing the company wanted, and they wouldn't hire me if they suspected I was going to cause one. And even if they were naïve enough to think a turf war would be ok, I knew I'd lose. To me, though, it would be worth it to get as much for users as I could. But I couldn't tell them that. Fortunately, I had come to grips with this whole issue back when I decided to become a UI Designer, so I was prepared with a plan for this critical moment in the interview.

My plan was to persuade the interviewers that the developers would be glad to let me manage their UI Design turf. I began by asserting that I understood the developer's mindset. I'd worked successfully with them as a technical writer for more than ten years, and I know where they're coming from. As evidence for this, I offered the highly positive references from developers on my resume.

As intended, these confident statements aroused the interviewers' curiosity, so I proceeded with some elaboration, as follows:

> *Developers are highly rational. They demand reasoned arguments and logical explanations before they take action. They bear the burden of the software project because they must write code that works without exceptions. Their code must work reliably all of the time. This is a heavy burden, and I respect it. As I've described earlier, the way I work always follows a clear, logical method, and the designs I provide show it. Developers appreciate that I have a sound*

method and that I can explain and justify it with good arguments. This builds their confidence in me. Finally, developers expect a UI Designer to understand their technical environment—to the extent that the designer will stay within the technical constraints of the current project and not come up with some blue-sky ideas that cannot be achieved within the current Tech parameters. This expectation is quite reasonable, and I try hard to meet it. That is why I always do my homework on the technical side when I start a project. For example, I ask what software platform and associated technology stack development will be used, and then I research that information so I can intelligently discuss with developers how my designs might be constrained and any limitations they foresee. When they see that I am knowledgeable enough to understand the relevant basics of the technical situation, they are happy to have these discussions and more likely to trust me with the design. I pretty much seal the deal with them when I present my first designs of the Conceptual Framework. That is when they see the proof in the pudding—my designs make sense, they are grounded in reliable user data that I can explain and justify, and they conform to the project's technical environment. Once I've won them over with the initial design, we can usually work together smoothly for the rest of the project. Of course, we may have some points of disagreement, but in my experience, we are able to compromise on a solution that works for both of us. It is important, of course, that development knows I have support from management. Will I have that support in this position?

By ending with this question, I put the interviewers on the spot. If they were persuaded by what I'd just said, they would feel compelled to answer that they will provide the support I need. If they actually gave me that answer, I would smile and nod, showing my relief and appreciation for their pledge. Of course, I was fairly sure that, when the chips were down, they would never go against their developers—no matter how much they might want to support me. But that was OK. I would take the chance. I would try my best to get along and see how much great design work I could accomplish for the users I'd be taking with me in my Tech car.

After this last, tricky discussion, the interviewers were finished with me. With smiles all around, I gathered up my props and left. The job interview was over.

At first, even though the interviews seemed to go well, I rarely heard back from them. But all it would take was one offer, so I cast a wide net and managed to get offers when I really needed a job. In a year or so, I was on my way with a design portfolio, making more and more money and getting some interesting UI design jobs. As it turned out, my approach to "getting along" with developers wasn't half bad. It worked for me more than once. Sometimes they really did let me drive on their turf, even though they always stayed in control of how far I could go.

I didn't care, as long as I could do UI design work and, maybe—just maybe—make software that satisfied and delighted users. To that end, I often found myself applying for jobs that included UI design work, even though the position was not explicitly looking for a designer. For many of the jobs I took, I was, in fact, masquerading, and I eventually tired of that. The time finally came when I didn't want to pretend anymore. The design of the UI was the UI Designer's turf, and I needed to claim it openly or go home. But before that realization happened, I wore whatever mask would get me into the party. Once in, I "wheeled and dealed" as much control of the UI design turf as *they* would give me.

CH 7. My Lonely Masquerade

Coming out of the Tech Bubble of 2001, I was encouraged to see that software companies were gaining interest in hiring someone who could help them increase user satisfaction. But as I read through the job descriptions, I rarely saw anyone looking for a UI Designer. I found that some companies saw user satisfaction as more of a customer engagement or customer hand-holding effort than a commitment to improving their software user interfaces. For such companies, their claims that "we listen to the customer" made good marketing but did not result in any change in the way their developers created the user interface. But at least they did carve out a role in their organization for some type of user advocate.

Fortunately, I did find some software companies that bought into the idea that improving their software's UI design was the way to increase user satisfaction. And these companies also believed that better UIs would come from a better understanding of their target users. Since their developers had no interest in user engagement, they decided someone had to be hired who could do this user satisfaction work. Someone had to be hired who could meet with users, learn their needs, and then communicate this information to developers. But hiring a "designer" to fill this role was often not the first, or even second, thought of many companies. They wanted someone who could understand their business and also the business of their users. They wanted someone who could put one foot in the user's world and one foot in the Tech world. They thought of designers as people who draw pictures, and that wasn't what they needed.

Even though a UI Designer really was what they needed, in those early days you couldn't blame software companies for thinking the opposite. The UI design discipline was still in its infancy, especially anywhere outside of Silicon Valley. Among those few newcomers, like me, who were trying to take on the UI Designer role, there were continuing discussions, disagreements, and competing proposals about the skills required for UI design and the methods that worked. At the same time, many well-respected usability professionals were still clinging to their turf, telling their managers that these new so-called UI designers were unskilled and would lead them down the wrong path. Since the usability folks couldn't design, they didn't want anyone else to either.

For example, I remember an interview I had with a major US defense company. Their Human Computer Interaction (HCI) research group was looking for a Human Factors Engineer who would work with their software teams on user interfaces. I interviewed with the manager and the senior usability professional. We were exploring the notion of a drone operated warship vs a human one. I argued that a drone ship with well-designed software that remote operators could use to fire on enemies was the better solution. The usability professional insisted that only a human who was physically present on the warship should be trusted with pressing the FIRE button. This well-published Ph.D. went on to say that putting effort into UI design and working with their development team on software projects for drones would not be fruitful compared with the value obtained from studying the psychology of human behavior on warships. I made her nervous, and she felt she had to put me down hard and fast in front of her manager.

When the manager called to tell me that I didn't get the job, she spent a while saying how much she wanted to hire me and start doing the work I had discussed with them, but in the end she couldn't go against the senior HCI person on her staff. I was disappointed, but I wasn't surprised. Most software companies were pretty much ignoring the idea of hiring a UI Designer, if such professionals even existed in the job force available to them. Companies looked at the roles they were familiar with and wondered which one of those could provide user satisfaction. If these companies had usability folks on staff, they would yell, "Me, me, me!" insisting that usability research was the way to user satisfaction. They were panicky at the idea of some upstart UI Designers replacing them.

I empathized with those usability folks, but I was now a UI Designer, and I knew these companies needed a UI Designer. But for the reasons just discussed, when I looked over the job openings, the title of UI Designer was rarely listed. So, on a whim, I started a more generalized search for any job descriptions that included words like "users" and "design" and "user requirements". Aha! I found a decent number of jobs that were, in fact, right up my alley. These jobs were asking for someone who could meet with users and gather user requirements. They were asking for someone who could communicate those requirements to developers and ensure that the software met those requirements.

Essentially, the companies with these job openings needed someone who could provide UI designs that would satisfy users, even though that is not what they were explicitly asking for. Maybe they didn't know it, but they needed exactly the skills that I now had. With the right spin on my resume, I could easily demonstrate to them that I qualified for their jobs. What a devilishly gleeful turn of events! I realized that I would be able to do more UI design work if I disguised my UI Designer title and billed myself as something else. The same resume would work for me; I just needed to change the title of my role and refocus my work history to show off the skills and experience related to user requirements gathering and communication of user needs to project teams. Basically, I would hide the UI Designer role behind a mask, even if UI design was the work I would be doing.

So it was that, in the early years of my design career, I could often be found doing UI Design work under the guise of one role or another. Sometimes my UI design work would be noticed for what it was and cause a little stir. When I was masquerading as a Business Analyst or a Project Manager, for example, someone would say, in surprise, "You mean these screen designs came from the BA (or PM)?" Yea, though I wrapped myself in the cloaks of the others, my UI Designer's robe always lay beneath.

My masquerade was not unique. More than one colleague I encountered in those days was doing good UI design work as part of a Business Analyst, Information Architect, Human Factors, or Project Manager role. With all this masquerading going on in those early days, in the rare case that a company did decide it needed a UI Designer, it was actually hard to find one by advertising for that job title. As a number of recruiters told me, the hiring manager would say to them, "How the heck do I find a UI Designer, anyway?"

7.1 Developers are from Pluto, but where do UI Designers come from?

Between 1996 and 2014, I held 17 jobs. Of those 17, 11 were masquerades. Four of the masquerades were as a Technical Writer/Documentation Manager (TW/MGR), 4 as a Business Analyst (BA), 2 as a Human Factors/Usability Specialist (HF), and 1 as a Project Manager (PM). My masquerades were successful because the skillset needed in my UI Designer role (as I described in the previous chapter) has much in common with the jobs I took in masquerade. The table below lays out the skillset crossovers more clearly.

	JOB TITLE				
SKILLSET	Tech Writer	Business Analyst	Human Factors	Project Manager	UI Designer
Understanding user needs	✓		✓		✓
Understanding user goals			✓		✓
User advocacy (always putting users first)			✓		✓
Communication	✓	✓		✓	✓
Schedule management				✓	✓
Meeting management		✓		✓	✓

SKILLSET	JOB TITLE				
	Tech Writer	Business Analyst	Human Factors	Project Manager	UI Designer
Requirements gathering		✓		✓	✓
Interfacing with developers	✓	✓		✓	✓
Documentation	✓	✓	✓		✓
Customer and user liaison/rep		✓		✓	✓
Technical knowledge	✓			✓	✓
Knowledge transfer between team and customer		✓		✓	✓
Executive presentations		✓		✓	✓
Stakeholders requirements management		✓		✓	✓
User acceptance testing			✓		✓
Requirements mapping to software development		✓		✓	✓
Schedule management		✓		✓	✓
User requirements	✓	✓	✓		✓
UI designs and prototypes					✓
Agile project experience	✓	✓		✓	✓

Recruiters recently sent me descriptions of job openings for a few of the job titles in the above table, and I show them in AP 5 on page 243. You can see the skills required and requested in those descriptions. Almost all of the listed skills are part of the UI Designer skillset. However, some of those desired UI Designer skills are unique to UI Designers and are not truly part of the skillsets of these positions themselves. That's why the recruiters sent me these jobs. They knew I had those UI Designer skills plus most, if not all, of the rest, and they thought I might be interested. Why not take a BA job if I can also do UI design as part of that role? And that's how my masquerade started.

By the way, those of you readers who are Technical Writers, BAs or PMs might notice from the table above that you already have many of the skills required for UI Design work, if you are ever interested in moving to that career field.

Of course, I had come from a technical writer and manager background and had worked in many different project environments, which grounded me in many of the skills listed in the table. The others I acquired as part of my self-training to become a UI Designer. While I wasn't an expert in every one of those skills, I was very good at the critical ones, and I was a quick learner for the rest. Most important, I was usually able to convince my managers to let me expand my role with users. They liked the caliber of my work, so when I asked to do even more, how could they say no? And once I was able to gather information directly from users and communicate it to the stakeholders, I almost always was allowed to move forward with suggestions for screen designs. Development often rejected them, but sometimes I prevailed and accomplished something good for users. These accomplishments I added to my portfolio so that, if a UI design job popped up, I would be first in line for consideration.

During my masquerading days I noticed a disturbing trend. As Tech was coming out of its bubble in 2003, the big Tech companies were the first to start hiring Cleanup Crew again, so those were the companies I looked to for opportunities. In my initial masquerades in the BA roles with these companies, I worked with the typical large teams of developers. But as time went on, I noticed that the teams were shrinking in an unusual way. I would walk into my first team meeting at a new job and there would be, at most, two developers sitting at the table. I would learn that they were the only developers from the company assigned to the project and that this was just one of the many project teams they were on. As for those other project teams, these same developers were the only company developers allotted to those teams as well. What was going on? "Where are all of the company's developers?" I would ask my teammates. "Oh," they would say, "haven't you heard?"

7.2 Congratulations, you've been out-sourced!

When the Bubble hit in 2000, large numbers of developer Lords had been laid off or resigned, fleeing to startups and game development. After the Bubble finally lifted, many tried to get back into big Tech, but few made it. The big companies weren't hiring them. Why? Well, to survive the Bubble and stay in business, Big Tech started looking to find new ways of working cheap. In short order, a wolf appeared at the door flashing a big, hungry grin. "My developers can do all of the work you need at half the cost," it growled. Big Tech couldn't resist and let the wolf in.

A harsh wind blew in with that wolf, sweeping away all of those fat, complacent software development groups that had ruled the roost at the big Tech companies, the big banks, and the giant insurance and pharma companies. In their place, organizations conceived and created in faraway and unfamiliar countries installed their contract workers. These offshore contractors became the new sets of developers assigned to a company's software projects. Hiring these unbelievably cheap, external, remote, culturally disparate organizations, and trusting them with the company's software development was called "outsourcing." Hey, with a deal like that, what could possibly go wrong? Hahahaha

It was such a cost-saving move that big Tech companies ran as fast as they could to get onboard. The old management guard was thrown out and new VPs, Senior VPs, and CEOs who had experience with outsourcing were quickly brought in to lay off the lordly developers and outsource their work. Oddly enough, many Cleanup Crew professionals were kept in place. They were assigned to the new contract developers and expected to perform their cleanup work as usual. But as a backup measure, most companies kept a few developer lords around to act as the overseers of the offshore group. Their role was to ensure that those "foreign contractors" were delivering acceptable code for the company's various software projects. On our side of the pond, the operation of this hybrid project team was alien territory. Welcome, strangers, to a strange land.

"Outsourcing" became a powerful word. In a Tech company that had not yet experienced the reality of it, merely uttering the word in passing caused a wave of fear and anxiety that rushed through the project teams. Who said it? Just one person? In a meeting or just in the hallway? When? Why? How? etc. etc. Developers who thought they had survived the Bubble were once again struggling with resumes and networking with colleagues. Members of the Cleanup Crew also panicked. The gossip around town said it was painful to work with outsourcing organizations. Numerous difficulties presented themselves on a daily basis as project teams interacted with these organizations. Some difficulties were minor, others major. But there were so many of them that the entire software development enterprise felt like it was built on sand. Here's a list of what I and my project teammates encountered while working on outsourced projects:

Outsourcing and language barriers

Most outsourcing organizations were from countries that spoke languages Americans rarely, if ever, heard or studied in school. The burden was on the outsourced organization to provide contractors who could speak English. This burden was typically not met satisfactorily. While many of the offshore contractors on a project team could read and write a fair amount of English, only a few of them could speak it so we could understand them. We were always asking them to repeat themselves. On their side, they usually could not understand us very well either, but even worse, they didn't say so. As a result, communication was haphazard at best, and this showed when the delivery of code came with unexpected issues.

Outsourcing and time differences

The contractors typically remained in their own country during a project. Sometimes several of them would visit the US team for a few weeks to solidify project planning, but this was infrequent, if it happened at all. Because of extreme time differences, most of the project meetings between the US team and the offshore contractors had to occur outside of normal US business hours, e.g. very late in the US workday (after 6pm) or very early (before 8:30am). This schedule did not allow for the rapid communication that software teams need when problems occur that require swift planning and reaction. It also forced US teams to meet when some of their critical members were unavailable. The result was ineffective team meetings that were not able to resolve problems promptly and efficiently. Confusion abounded over who was doing what when.

Outsourcing and infrastructure disruptions

Connectivity between the US and the remote sites where the contractors worked was often unreliable. In the early days of outsourcing, most connections were simple, relying only on conference call connections. Even so, the subpar telecom infrastructure at the remote sites often produced unacceptable background noise on speakerphone. The noise, combined with the broken English of the non-native speakers and the slow delivery of their remote connection, made these calls frustrating at best. People were either waiting forever to hear a response or jumping in and talking over someone. Worse than that was the occasional drop of the call altogether, which resulted in both sides trying to redial and running over each other or just giving up.

When web conferencing became a reliable and fast alternative in the US, it was also tried for these remote meetings. The connections could be established, but the data transfer was almost always unacceptably slow. Screens would be frozen for too long while the team waited for the data to transfer. People would walk out of the meeting because they just couldn't waste any more work time on such an inefficient process. Companies soon learned that the connections would never work properly unless the US company bit the bullet and outfitted the remote site with US equipment and paid heavy fees for better connectivity.

Outsourcing and cultural disparities

The cultural differences between the US teams and the outsourcing organizations were often major. This was particularly true of the way women team members were treated. US team members soon realized that women on the remote teams were often restricted from answering questions and making decisions. They always deferred to one of the men on their team. As for women on the US teams, we found that our instructions to a remote team and requests for information from them were sometimes either ignored or pursued with one of our male US team members. This was not an issue of poor manners but rather of doing business amid disparate social structures and values. While we in the

US tried to accommodate the diversity, some differences could not be accepted as OK. No one on the US teams was comfortable with this situation.

Another cultural sticking point was the level of commitment to the delivery schedule. If the US team got behind, we would arrange vacation time so work could continue over major holidays—at least enough to get the schedule close to back on track. Also, US holiday periods usually lasted a few days and never ran longer than 10 business days.

In contrast, many of the outsourcing organizations strictly observed one or more holidays that lasted for 3-4 weeks. No work could be performed by them during these periods. If the schedule was running behind, there was no recourse but to wait for these holidays to pass. After they did pass, the remote contractors were willing to work overtime for many weeks, but the US team members could not conform to that type of schedule.

Neither group is right or wrong here, of course. Cultures differ and that needs to be respected. But the consequences of doing business amidst those differences can be severe, and in my experience, they often were. Deliveries were not met or else the quality of the delivery was unacceptable and full of unexpected issues. Most frustrating was that both US and remote teams ended up pointing fingers at each other as the reason for any project failures.

Outsourcing and code quality issues

The technical changeover from US development teams to outsourced development teams was jarring. In my view, the change has never achieved the success it promised. How could it? US companies were replacing creative developers, who sought elegant code through innovation, with programmers of limited skill and experience, whose bosses demanded quick and easy code while rejecting efforts that would sacrifice time for better quality. Coding jobs in offshore companies were extremely highly valued by the citizens of those faraway countries. Those locals who were lucky enough to be hired were not about to rock the boat. They went along with whatever their bosses demanded. They feared displeasing their bosses more than they feared delivering what the US team considered to be sub-standard code. As a result, sub-standard code was typically delivered more than once during a project.

For example, I saw problems with poor performance and tangled navigation in the UI. Searches did not deliver the expected results. Buttons and links didn't work or went to the wrong screen. More than once I heard developers complain that the code was missing comments. It was also typical to see the text of screen messages inserted directly into the program code instead of separated into an external text file. That meant every change to a screen message—even to correct a spelling error—required recompiling the code and redeploying it.

Dynamic problem solving became another quality issue. The developers on the offshore team seemed like one-trick coders who insisted on coloring within the lines. When code needed to be extended or redesigned to accommodate a new feature or the enhancement of an existing feature, the suggestion of writing custom code to solve the problem was rejected. If the problem could not be solved using the basic code constructions shipped with the current programming technology, then the problem couldn't be solved. Custom code, often called "NOT out of the box" code, was not part of the deal. US companies either had to hire their own developers to write new code or else give up adding those new features and enhancements and live with the code that was delivered. Most of the time, they lived with the code that was delivered. As a result, the criteria for acceptable code deliverables often sank to sub-standard levels. Sadly, even now, many years after the dawn of Tech's outsourcing debacle, we continue to experience the effects of these new "standards."

Outsourcing and UI design issues

The appearance and flow of UIs are still suffering from the effects of outsourcing. Usability issues that we thought were overcome in the late 90's resurfaced with outsourcing (see the list of Common Usability Issues in CH 5, on page 107). User satisfaction as a critical goal was foreign to the offshore teams, and it has only lately been added to their approach in a perfunctory way. They placed no value on understanding how target users think, and even now they have little familiarity with how to work with users towards this understanding.

In this outsourcing milieu, achieving my goals for users became complicated in ways that I hadn't experienced. I was used to the haughty Lords who could create elegant code, reconfigure projects to come up with exciting innovations, and, if they were willing to accept my UI designs, also deliver software that

delighted their target users. Now, they were gone, and I was left with code monkeys who had no interest in users, elegant code, or innovation. Here is one example of what I'm trying to describe.
In 2014 I was the BA on a bioscience project that used an offshore development team. To my delight, I was put in charge of designing screens for the software. In particular, I was working on a series of online forms that took the user through a strict sequential flow. Input had to occur in the correct order. The user could go back to previous screens and make changes, but if those changes affected the input already completed on other screens, we needed to make it easy for the user to access and update those other screens as well.

One of my major requirements was that the user's input would be automatically saved on a screen as it was entered. Users would not have to worry about losing data if they jumped to a previous screen to make a change or jumped to another menu function in the software, or even if the system crashed. All data would be auto saved as it was entered or modified. The developers from the offshore company did not want to implement this requirement. During a team meeting they insisted that a Save button should be put on each screen instead. "And if the user forgets to click the save button?" I asked. Their response to that question was, "Too bad. The data will be lost. They have to learn that they need to save their data with an explicit action. We can't be saving data on a screen before the user completes everything on the screen and keeping track of all that data if the user decides to jump around to a different screen."

Uh-huh. Of course, they could auto-save. They just didn't want to, or maybe they didn't know how to do it properly. I told them my requirement was critical for users. Someone suggested that a message reminding the user to save should pop up if the user tried to go to another screen without clicking the save button. I insisted that would just interrupt users unnecessarily and require them to complete more actions than were needed. It put an unnecessary burden on users.

"Clearly," I stated at the project meeting, "users want their data saved as they proceed through the screens to complete their task. And even if they have to leave a screen before it's completed, they want the data saved that they have already entered. Furthermore, if the system crashes while they are on a screen, it's standard practice to save that data for them. Obviously, we should also apply that mechanism to the data while users are entering it." I reminded the team that their iPhones always saved their data after they entered it. "How many save actions do you do on your phone's message apps and notes apps when you jump away from them to answer a call?" I asked, rhetorically. The offshore developers grudgingly agreed to auto save on each screen.

At our team meeting some weeks later, they demonstrated their code to show what they had implemented. The screens had Save buttons. There was no auto save. When I asked the developers why the Save buttons were there, the US project manager jumped in and said that he and the offshore developers had discussed it. In the interest of time and avoiding coding difficulty, they decided to use the Save buttons instead of the auto save. I walked out of that meeting, turned in my company laptop and badge, and left the building for the last time.

I later heard from my manager, who left a few months after me, that the offshore team was so flummoxed by my abrupt exit that they went with all the rest of my designs without question. Of course, I'm sure they still used those user-torturing Save buttons.

Outsourcing remediation

On several of my projects, I was able to work effectively with remote teams, but only if some of the major hurdles listed above were overcome. The solution always lay in the use of adequate technology by both sides, agreed upon code quality standards, and the demonstrated ability of offshore contractors to communicate effectively in English. These things were needed to ensure that the project team could respond to issues, questions, and problems in a timely manner. Without a doubt, I had the privilege of working with some great developers from offshore companies. They definitely broke the mold and typically ended up being hired by US companies. Nonetheless, outsourcing carried so many risks that everyone on the Cleanup Crew dreaded the loss of US developers and the changeover to offshore companies.

As the Cleanup Crew shuffled along under the "do more with less" work ethic, trying to clean up after sub-standard deliverables, a whole new skill was getting the attention of US companies: Outsourcing Management. Oboy, I was all over that like white on rice. I had already decided that it was critical to work well with remote teams. I could see that outsourcing was not going away, so I knew that my success depended on adapting to the new world. Fortunately, my

academic stint as a university professor had given me the tools I needed for what I thought of as "outsourcing remediation".
I had taught many culturally diverse students with serious language barriers and had worked alongside academic colleagues from faraway lands. Two communication techniques I learned from those days helped me again with offshore teams. I remembered these techniques as I noticed the communication problems team members were having during project meetings. The underlying causes for most of these problems were:

- US team members failed to appreciate the importance of speaking slowly to non-native speakers so they had time to process the unfamiliar English vocabulary and sentence construction.

- If non-native speakers on the offshore teams didn't understand what was just said, they were very reluctant to stop a conversation and ask for repetition. Rather than display their language difficulties to the team, they would wait till the meeting was over and hope their English colleagues understood the conversation enough to fill in the blanks.

- US colleagues failed to stop frequently and verify that the offshore team understood what had been said. Verification requires the offshore team members to repeat, in their own words, what's been said to them. Instead, US colleagues typically took a "Yes" as sufficient verification or else decided the burden of understanding was on the offshore team, so it was up to them to ask for clarification.

I recognized these causes of communication failures from my academic days and remembered how to fix them. I knew how to set the tone with remote teams so that they felt at ease asking for repetition and clarification. I also knew how to stop at critical intervals to verify that communication had been successful. For example, when it was my time to speak at our first team meeting or if I were facilitating the meeting, I would begin by saying something like this, somewhat slowly, to our offshore team members:

> *I would like to talk to the team about communication. Often, we native English speakers speak only English, and sometimes we forget that the whole world doesn't speak*

> *English. (This usually got a chuckle from everyone.) I know you in our offshore group may speak several languages, and we over here don't expect you to speak perfect English. So, please understand that we appreciate that you are willing to speak our language for this project, even though it is not your native language.*
>
> *I will try to speak somewhat slowly, and I will repeat the things that are very important. But I need your help. The most important thing to you over there and to us over here is to work together to do a good job. If we over here don't understand something you say, we will stop and ask you to explain it. If you over there don't understand something or you would like us to repeat something, please ask us to explain or repeat, OK?*

Then I would wait for their response. If they understood what I'd been saying, they would usually respond with something like "Yes, we will ask questions if we don't understand." If they didn't understand what I said, they would usually either admit that or just say "Yes." In either case I would then prompt them with questions to see what they really understood, if anything. This would help me gauge how bad the language barrier was going to be.

Even in the worst cases, setting this kind of no-fault tone made it possible for us and them to ensure that we had effective communication, even if we had to proceed very slowly. Often our communication improved at each meeting because both sides were comfortable asking for clarification and verifying that each had the correct understanding. Sometimes there was good-natured laughter by one side or the other when a team member stumbled with a language oddity or made a blooper.

As I stepped up to manage communication this way on my projects, the US team members quickly started to show their appreciation by making me the primary go-between for team communication. They were happy to have someone on board who could handle the language barrier, since it had so often been a serious point of failure for outsourced projects. They gave me so much control that I was able to enforce a phone etiquette that avoided interruptions and allowed each member adequate time to speak.

Everyone appreciates a well-run meeting—especially with offshore teams—and I built on this experience to advance my marketability. Soon, with a bit of resume massage, I was able to add an Outsourcing Specialist mask to my repertoire. As it turned out, this mask later gave me entry into one of my most productive and enjoyable UI Design jobs. Here's how that happened.

A large retail software company was advertising for a Human Factors and Usability Engineer. My first phone interview for the position was with the development manager. He didn't spend time discussing my Human Factors and UI Design experience. Something else was on his mind, and he got right to it. He said he had a problem that he needed someone to solve for him. His team was developing a new, highly graphical, UI for their retail software, and he'd been required to outsource the graphics. Unfortunately, he was finding the offshore team difficult to impossible to work with. The language barrier and time differences were creating havoc with the deliverables from the offshore contractors. No matter how many teleconferences this manager had with them, they delivered the wrong graphics, or unneeded duplicates, or failed to make the requested updates, or failed to deliver on the scheduled date. He said he just didn't have the time to deal with it anymore, and this problem was ruining his delivery schedule. He asked me if I could I fix it and how I would do it.

Yeah, I got the job, I fixed his problem, and the rest was history. My title for that job—essentially, UI Designer—was what I really wanted. And UI design work was what I ended up doing for the company for 3 years. However, to get to that place, I had to start out wearing my Outsourcing Specialist mask. I was OK with that then. Now, though, I wouldn't want to spend my time fixing such problems.

In any case, throughout the Outsourcing era, I danced along with various masks: Business Analyst, Project Manager, User Acceptance Tester, Technical Writer. Those masquerades gave me the chance to do some wonderful UI design work that actually saw the light of day. And I met some heroes along the way. Here are some of those experiences.

7.3 Masquerade at a bank

For this project I wore my BA mask. I was placed on a team that was responsible for reworking the current branch banking account management software. This software was very out of date. It still used the old mainframe green screen application, even though it now displayed on up to date desktop machines. As with all the old mainframe green screens, the bank account reps had to use F keys to navigate the screen and activate functions. And, typically, each screen had limited functionality, so performing a task often required a sequence of 10 screens. Of course, the bank reps could not go back to see what they had entered on previous screens, so it was easy to make mistakes—and they were only notified of a mistake after they had worked through the entire sequence of screens!

Our project team's goal was to replace this terrible old mainframish application with a PC software client that offered a graphical, mouse-driven UI, a better flow through user tasks, and intuitive screens that wouldn't require extensive training to use. The team I joined had not yet experienced outsourcing. It contained a team manager, a small development group with its own development manager, a few testers, and a small usability group. It was 2005, and the big banks were some of the first companies to care about the user interface, so I was looking forward to the work.

The team manager worked closely with the development manager. Although she was technically in charge of the entire team, she and the development manager worked together as equals. The great thing about this team was that both of these managers were on the same page about the importance of a well-designed UI. The development manager was one of the few good guys I worked with during my masquerading days. He was ready to have his group code what the designers came up with, as long as it made good sense to him. The folks on the usability team, on the other hand, were clueless.

This became evident at the first team meeting I attended. Witnessing their ineptitude, I understood why both of the managers had brought me on. After 6 months of work, the usability group had come up with a set of screen mockups for the new UI design. Having spent a week studying the old green screen application, I was now ready to see this new design. As they went through their screens at the team meeting, it was painfully obvious that they had simply mapped the functionality and screen elements of the old application into graphical PC screens and christened it the "new design." They had even included the same mainframe F keys as the mode of interaction with the screens! Their screens required user tasks to be performed just like they were performed with the old application, including the same sequences and the same number of screens in each sequence. Although their new designs did allow users to go backwards, a user had to click back through the screens in sequence to get to the screen they needed to see. So, if you were on the tenth screen but needed to see something you typed on the second screen, you had to click back through screens 9 through 3 to get to screen 2. What a joke!

When their presentation was finished, I didn't hold back. I pointed out that the new screens were a direct mapping of the old screens and, hence, the new design perpetuated all the usability issues of the old design. There was shocked silence in the room, and then both of the managers spoke up and agreed with my observation. They asked me and the other designers to go back to the drawing board and come up with a better design. As far as I could tell, the managers had hired me for two reasons: first, they wanted someone with design *bona fides* to say out loud what they had been thinking about this ridiculous new design; and second, they wanted new blood who could come up with something better and also engage the usability group to work on it. Oh yeah, I was their gal.

I asked for a meeting of the usability group and the team manager. At that meeting, I explained my design methodology, emphasizing the need to get out and meet with the branch bankers to capture their current, most pressing software issues and their key needs for the new software (sound familiar?). After we captured that information, I explained, we would be able to come up with a design that would satisfy them. The team manager loved the idea, the usability folks not so much. The 3 of them were contractors, like me. But, unlike me, their backgrounds were designing eye candy on the bank's web pages and doing usability testing. They had no idea how to approach real UI design. Oh well, they were in the soup now.

As the first two weeks went by and we prepared for user meetings with the branch banking reps, one of the usability group's contractors left for a graphics job at another bank. Of the remaining two, one was fairly sheepish and went along with the flow. The other one, who had been the leader of the first, disastrous design effort, hated my guts and fought my ideas at every turn. She started by marching down to the team manager's cube and telling her that my ideas would never work and would be a waste of time, especially since we already had her design ready to go. The team manager told her that we all should give my ideas a chance, and the development manager even chimed in from across the aisle with his full support of my ideas! Hearing all this from my cubicle down the aisle, I saw the angry "designer" turn white. She had thought the development manager was her best buddy, based on all of the joking banter and kibbitzing they did together during breaks.

Now, realizing that she had no support to stop me, she decided to spend lots of time arguing with me about my methodology instead of doing the work we needed done. Each day, before I barely sat down at my desk, she started going at me. I argued with her kindly, hoping for persuasion rather than bitter victory. But I didn't let her slow me down; I continued to move forward with my method. She stayed at the office while I and the other contractor in the group traveled to the branch banks and held meetings with the bank reps. On the ride back to our offices after our last meeting, that colleague told me what a great experience the user meetings were, and that he was now a believer in my methodology. So, onward we went, dragging the angry, whining, stubborn "designer" lady through my process.

The major idea I came up with for the Conceptual Design of the UI was a tracking section on each screen that would show the user what steps had been completed in a task sequence and what steps remained to be completed. In my design, the Tracker, as the development manager joyfully christened it, would always be available, and users could go back to a previous step by directly selecting that step from the tracker. I mocked up the Conceptual Design showing the Tracker functionality and presented it at the team meeting. The managers loved it and loudly demonstrated their approval. My supportive colleague was happy to be part of a good thing. The angry "designer" stated that the tracker would never work, even though she couldn't give any reasons why. All she had was "It's too complicated. Users won't like it." My response was that we will bring a demo to them and see what they say. She rolled her eyes. The managers both said thanks, this is wonderful, keep going, build out the rest of the screens.

I did keep going, although the angry "designer" was beginning to weigh on me. When I arrived in the morning, she'd either be at my desk ready to fight about her design vs mine, or else at another cubicle, loudly talking trash about me to someone else. Why didn't she just quit? Why didn't they fire her? I'll never know. To this day I haven't figured it out. In any case, I finished the screen designs, the bank reps loved the high-fidelity prototype I made for them to try out, the bank gave me an award, and the developers started coding my design. At that point I got a call from a recruiter who had an even better BA job available in a pharma company. At the same time, the VP at my current bank job was starting to outsource development, and there were rumbles that our team was not safe. So, I jumped ship and took the other job.

About 7 months after I left the bank, the team manager called me with the news that the bank started deploying the software with my UI design, including the Tracker, and so far, the branch banking reps loved it. She thought I'd want to know. That was a great day for me. Sometime after that I heard from a former bank colleague that the angry "designer" had left the bank months ago. Her job had been outsourced.

7.4 Masquerades at pharma companies

Pharma mask 1

Wearing my Usability Specialist mask in 2002, I took a contract with a multinational pharma company that included a few trips to sites in Europe. I worked with development to provide screens for a wizard-style scientific search application for internal company users. The development group's manager gave me free rein to talk to users across multiple sites. I was able to follow my methodology, and the developers, most of whom were remotely located, were supportive and highly skilled.

I finished the design, and the developers created a prototype. I was able to verify the prototype with a small sample of users. The user feedback was highly positive. The development team presented it to management and received high praise for the work. My job was complete, and I went on to another gig.

A month later, one of the developers in the group told me that their manager decided to "go in a different direction," and the project ended with only the prototype. A few years after that, I found out that the manager was attending conferences and showing the prototype as one of the latest and greatest projects going on in that pharma company's scientific software group. I guess our work was good enough to show off but not good enough to bring to fulfillment. Go figure. Big Pharma politics, I guess.

Pharma mask 2

Switching to my BA mask in 2005, I took a contract at another big pharma company. This was in the twilight of the Golden Age of Big Pharma, and their IT departments were still dripping with money. At the same time, the idea of outsourcing their IT needs was also starting to look good to them. I have to say, this was one of the craziest projects I've ever taken on.

The IT Senior Vice President (SVP) had spent a million dollars or more on a social media application that had been customized for the pharma company by the vendor. This customized product was supposed to provide sharing of scientific expertise within the pharma company. For this sharing application to work, the company's scientists would each have to complete the application's online form that detailed their areas of expertise. Once all of the data was in the application, it would be easy for a scientist to find others in the company who had the expertise needed for the scientist's specific research work. Essentially, the application would enable instant consultations among scientists!

That was the vision, and the application was deployed and introduced to the company departments with much fanfare and congratulations. Unfortunately—and to the SVP's surprise—the reality turned out to be bad news. Eighty-five percent of the company's scientists failed to complete the online form in the application. This would not do! The application had been purchased and deployed, and now the users were expected to embrace it. What was the problem? There was a small IT group dedicated to getting the application up and running, but despite their best efforts—internal advertising, brochures, demos, brown bag lunchtime PowerPoint shows—only a few scientists bothered to complete the online form in the application. After a year of failing to get at least 50% of scientists to participate, the small IT group's manager decided to hire a contract BA who would be dedicated to nothing else but fixing this problem. Enter the dragon lady.

I took the contract because I figured the problem was likely the UI, so maybe I could wiggle my way into doing some good design work. Plus, the money was exceptional, not to mention the delicious coffee bar in the main lobby. Well, I was certainly right about the UI. Any users who had attempted to complete that wretched online form were very likely going to give up in frustration and even anger before ever finishing it. In fact, I was amazed that about 20% of users had actually suffered through the effort and completed the form.

That UI was terrible in so many ways. Confusing, unintuitive, error-prone, and overall an unsatisfying experience. Users couldn't understand what they were supposed to do, the sequence in which they were supposed to do it, the relevance and purpose of the areas they were supposed to fill in, how the information they filled in was being used, the meaning of the display that was returned when a submit button was selected, and on and on. But as bad as that UI was, the application had a worse problem that was hidden deep inside its code. Finding that out was one of the most interesting experiences I've had in the software business, seasoned with a little bit of crazy.

Of course, there were two parts to this software application. One part was the population of the database with information about the expertise of scientists in the company. This expertise data had to be put into the database before searching would be useful. The target for populating the database was getting completed input from 70-80% of scientists in the company. Once database fulfillment was complete, the second part of the application could be used. That was the search function, which provided the real benefit of the application by finding the relevant experts in the database. When I began the project, I was told to focus on working with the user base of scientists to get the database populated. Little did I know at the start that the search function of this application would never be able to deliver the results it promised.

I started the project thinking that the right UI would solve the database fulfillment issue. When I later found out that, even if the database fulfillment succeeded, the search function would not be useful, I found myself on the horns of a crazy dilemma. On the one hand, should I try to fix the UI to entice users to complete the online forms for populating the database, knowing that they would eventually realize their efforts were for nothing? Or, on the other hand, should I save users from a bad application by reporting back to the IT manager that my investigation has revealed that the search function was not going to work? Which of these horns would be better for users?

I eventually decided to try to do both: fix the UI and let management know that the search function needed to be reworked by the vendor. That decision allowed me to continue in good conscience with improving the UI. The fate of the search function was out of my control, but I would do what I could to make it right. If it did get fixed, my improved UI would be ready. If it didn't get fixed, I was not responsible for the failure. In any case, the crazy part of this story is how I came to learn about the search function's problems.

The primary task the group manager had given me was to contact scientists in each department to find out why they had not yet completed their online form in the application and to reiterate to them the importance of doing so. To persuade them to "do their duty," I was supposed to explain the value the application to them and to the company as a whole. These were essentially sales cold-calls to users who had already decided not to participate, despite the strong encouragement from their own management, who themselves were give mandatory participation quotas from their SVP's. Contacting these busy, highly-intelligent, resistant (and annoyed) users would not lead to convivial conversations, and it was soon clear to me that I had been hired primarily to perform this unpleasant task so my manager and her group could avoid it.

No problem. I love talking to users and listening to their frustrations with their software. That's how I learn what will make the software better. I had reviewed the application before I made these calls. I knew it was a usability nightmare, and I was ready to agree with users and ask for their help to make it better. This attitude worked well for my first 50 or so calls. Users actually became gracious on the phone. They agreed that the application could have significant positive value if the online data entry forms could be improved and streamlined. All seemed to be going well. But then, I called the Russian.

He turned out to be one of my all-time developer heroes, and I still remember a lot of our conversation. He could easily have just told me he was busy and had no time to discuss the application. As a high-flying, major pharma research statistician, he *was* busy. But something clicked between us at the start of the conversation. Perhaps, as I plowed through my reason for the call, he sensed that I really did care about users. Maybe he thought I had enough of a brain to make the discussion worth a try. Whatever. He told me that he had tried the application and it didn't work. I responded that I wasn't surprised. Most users found the UI for entering their expertise data frustrating and hard to use. That problem was the reason for my call.

He chuckled while noting that, of course that UI was ridiculous, but that could be fixed. He told me that the application had a worse problem which, very likely, could not be fixed. That problem, he said, was the search function. I asked for elaboration. He gave me a real-world example. He said a colleague had filled out the online form and asked him to do a search in the application by entering one of the colleague's expertise areas as the search term. The Russian I was talking to (I like to call him "my Russian") entered a well-known synonym for his colleague's area of expertise and clicked the Search button. His colleague's name was not returned in the search results. Even worse, names of scientists were returned who were known *not* to have that area of expertise!

I was stunned. My Russian chuckled again, pleased that I appreciated the seriousness of what he'd told me. I knew the database didn't have much information in it, but I also knew that couldn't explain the odd results my Russian had received from his search. I said something like well, the software must not have an adequate dictionary or else its synonym matching is not very good. He liked that response and launched into what he knew about that subject and what he'd found out about how the application did its matching. Basically, my Russian had done his research on this awful application. He said the application was an unsophisticated joke, completely lacking the intelligence required to deliver what it promised.

I wasn't offended or defensive when he said this. I needed to understand what was wrong with this damned application, and lucky for me, I was talking to someone who could tell me. I asked for further explanation, which really got him going. Now we were touching the edges of the stuff he thrived on, and he seemed happy to educate me. As he went on, I started to get interested in that Tech itself. I knew I would never understand enough of what he said to bring it back in detail to my manager. But I still wanted to learn as much as I could.

Soon, we were talking beyond the realm of the problems with the specific application of concern. The discussion was basically off the books and off the record. We were talking for the fun of the subject. I appreciated him for taking the time, and he appreciated me for seeing the beauty of the problem. So, I learned a lot about the role of statistics in software search functions and the problems of synonyms and context and dictionaries, and the vagaries of pattern matching. I knew the whole conversation had taken a crazy left turn away from what I was supposed to be doing, but it was so fantastic to have this incredible expert teaching me about cutting edge Tech—giving me, with my limited knowledge, what could be just a glimpse into that realm.

We must have talked for close to 45 minutes. I thanked him for his help, and I never spoke to him again. Thankfully, he had given me the critical information I needed to let my manager know that the search function was a serious problem. I went to the manager and described the search failure that my Russian had found when searching for his colleague. My manager and the group then spent some time doing some test searches on the minimal data that was in the application. Numerous problems arose, similar to my Russian's. Clearly, Houston, we had a problem.

The manager discussed the search problems with the vendor, who dismissed them as outliers but promised to "look into them." I finished my report on what I had accomplished, complete with a review of the feedback I'd received from my calls to users. I included new screen designs of the UI that would fix the most important and common problems users had reported to me. The manager loved the report, and it looked like she was going to offer me an extension on the contract so I could continue working on that terrible application. I jumped ship as quickly as I could. I heard soon after from someone in the group who told me my UI changes were scheduled to be put into the application. Several months after that, when I took a contract with the same pharma company but in a different group, I learned that the application had been withdrawn from the system, and the manager of the group had received an award for forward thinking ideas. Cool. I would not have expected anything less.

But the craziest part was that a few years later, I had an accidental and brief encounter with the manager of the research statistics group that my Russian worked in. We were riding in the same cab and got to talking. The conversation I'd had with my Russian came up. The manager was surprised at how happy I seemed about that experience. He started smiling and told me that my Russian had a reputation for being unpleasantly straightforward in general, as well as rudely dismissive of people who expressed faulty scientific ideas. I couldn't believe it! My Russian? One of the most interesting, engaging, brilliant, and patient developers I'd ever encountered? "You must be thinking of a different Russian in your group!" I proclaimed. But it wasn't a different guy. For sure, I knew that my Russian was a great guy. He must have been misunderstood, misperceived, unfairly judged. Unless—maybe he was masquerading too, and had taken his mask off just for me, for those few minutes of our call.

Pharma mask 3

For my third pharma contract I wore a Project Manager mask. This was my first official appearance as a PM. At my interview, I talked around my lack of official PM experience by emphasizing the unofficial PM work I'd done as a documentation manager, UI Designer, and BA. I was clicking with the hiring manager, but she was a tough cookie. I could sense her skepticism about my PM *bona fides*. After waiting a few weeks to hear back from her, my recruiter told me that someone else had been hired. Christmas of 2005 was a week away, so I relaxed and enjoyed the holidays. Companies were not likely to do any hiring until January. But just before the New Year, the recruiter called to tell me I had the job if I wanted it. The first hire had stayed for two weeks, then left to take another job.

Hmmm, that was odd, I thought. Burning bridges like that was unusual for contractors, and it raised a red flag. A very early exit by a contractor often indicated that the job was bad news. But the money for this gig was great, and I'd liked the hiring manager and the nature of the work as she'd described it, so I decided to try it out. When I showed up for work in January, the manager didn't offer much explanation except to say that there were no hard feelings about losing the first guy, and she was glad to have me on board. I didn't like being the second choice, but as it turned out, this contract became one of my most rewarding UI design experiences. Life is funny sometimes.

The first project I managed for this contract called for a revamp of some old software currently used by a group of scientists to populate and access a database of experimental data. The tricky part was that the data they needed was stored in an old format in a legacy database, and in order to use that data, they had to first transfer it to a different database in a different format. Their old transfer application was slow, cumbersome, error-prone, and frustrating to use. Scientists found themselves lost in long sequences of dialogue boxes. Information was spread across the screens so users would forget the numbers of the data rows they were transferring as they finally got to the last dialog boxes in the sequence. They ended up writing large chunks information on paper as they proceeded through these sequences so they could retype it as needed to proceed. It was a dreadful UI, and they were very anxious to get an upgraded application.

To my happy surprise, the two developers running the project were top notch. The group manager was a recent company hire—a hotshot who had relocated from Silicon Valley. The team lead was also a permanent employee and a Java guru. The remaining 4 guys were contractors working on site. All of them were Agile aficionados, and I was up to speed on that development methodology from previous contracts. So, we were all *sympatico*.

As PM, I had to track the project's progress and enter all project information and updates into a new, cutting-edge Agile PM software tool. The manager was initially a bit worried that I would not be able to master this fancy new tool. Yeah, right. For me it was a piece of cake. As a favorite colleague once joked, "I'm not worried about coming up to speed on a new tool. I've already taught myself how to use 50 new tools, so I think I can teach myself 51." It didn't take long for the manager to be impressed with my PM prowess.

But I was also assigned other tasks that went beyond the customary PM role, and those tasks were the ones that lit my fire. The target user base of scientists had appointed someone in their group to represent them and work directly with our team to give feedback and answer any questions we had. Using agile speak, we referred to him as "the customer". It was my job to be the liaison with this customer. As liaison, I had to understand what the customer needed and communicate that to the team in language they could code from. This was perfect. I could apply my UI design method, at least to some degree! I was in heaven. But even better than the liaison communication part of my job, I was also supposed to ensure that the customer was happy with the UI of the new application. Gaining this much control over the UI went beyond my wildest dreams. About a month into the project, I exerted this control forcefully.

I had provided screen designs to the team based on my work with the customer. One of the customer's key needs was to have one screen that showed all of the relevant information needed to perform a data transfer. If we could do this for the scientists, it would remove most of the pain they were having with the old system, which spread the relevant information across multiple screens. My screen designs, which the customer had approved, satisfied this key need. The development group had accepted my screens and were set to deliver working versions for testing in a week. But when the delivery day came and I reviewed the working screens with the group, I saw that the contract developers had changed the design.

They had created two screens of relevant information instead of putting all of it on one screen, *per* my original design. When I questioned this at our review, they told the group that it was very difficult to put one of the pieces on the same screen. It would require a set of very difficult code, and they felt it was too much effort for the benefit it would provide. I had seen this crap before from the Lords of Teach and been forced to swallow it. Now I was the PM, and it was not going to stand.

I responded with something like, "So, you all have decided that the benefit to the customer was not worth the coding effort, even though the customer told us that this benefit is key to satisfying the needs of the scientists he represents. Is that right?" They looked at me like I had just crossed a red line. Did she just dare to say that to us? We stood there glaring at each other, preparing for a fight. And then, before everything went to hell, a hero emerged. It wasn't the manager, but the other permanent employee—the team lead, who typically hung toward the back of the room and never said much. In fact, to this point I'd had very little direct interaction with him and couldn't recall his name. He stepped forward and said, "The customer told us he needs this, so we are going to do it."

I was in shock, as were the contract developers. At least we had that in common. The meeting ended right there, and the software was reworked to support my original screen design. That project finished on time, and the application was deployed to the scientists. About a week after deployment, I received an email from the customer, copied to the development manager and the hero team lead guy. To my eternal regret, I did not keep a copy of that email, but I'll never forget it. It was short and sweet, and went like this:

> *Dear Eva,*
>
> *I wanted to let you know that the new software is getting rave reviews from my colleagues. We are so used to getting software we don't want and software we can't use. This is the first time I have worked with a development group. I was skeptical about the process. Now I can say that your team has given us the software we needed and gone beyond. Thanks for letting me be a part of it.*

Yes, I was masquerading, but this is what my heart sought, and achieving it made the mask tolerable. Reflecting on the experience later, however, I realized that it would never have happened without that developer hero. Even as a PM, I couldn't control the UI turf by myself.

7.5 Masquerades in government

Government mask 1

In 2007 I took a 3-month BA contract with an auditing group of a state treasury department. This group had been laboring under the weight of paper processing and desperately needed to automate their processes with workflow software. Their records room looked like a hoarder's living room. Going in there to find a file became a task to fear and avoid.

My job was to understand their current paper-based workflow and provide an analysis of how to eliminate the paper with workflow software. This analysis would then be given to the state's IT group, and they would provide a workflow application that satisfied the requirements and recommendations in my report. In the Tech business, this BA work is called analyzing the AS IS process and providing a TO BE process that solves the problems with the AS IS process. This analysis is typically done with flowcharts and other diagrams that show how the user's tasks flow from step to step in each process.

I put on my BA mask, but the UI Designer underneath was excited about the TO BE part of the contract. Maybe, I thought, I'll be able to include screen designs in my TO BE report. That way I could flesh out the flowcharts with specific ideas of how the software should look to ensure it solves the user problems. The AS IS part of the work was looking good, too. I would be working onsite, which would give me easy access to the users (who were all experienced CPAs). I would be able to observe each step in their paper processes as they passed documents from one desk to the next. Everything looked positive for this contract, but I had an unexpectedly rough start.

As I began working with the accountants, sitting with each of them, observing each one's activities in the paper workflow, and discussing their specific goals and problems, I ran into blank walls. No one really wanted to talk to me. I sensed fear and frustration among them. The misery they felt working with their paper process seemed to be outweighed by the fear of having their jobs replaced by new automation software. It was also obvious to me that they believed that automated workflow could never successfully replace skilled accountants. These CPAs were obsessed with accuracy, to a degree that I'd never seen before. They were software savvy and fully appreciated its capabilities, but they didn't trust it to make the right decisions. In their realm, decisions about financial solvency and the trustworthiness of borrowers needed to be made by the professionals. The more I learned about their work, the more I agreed with them. In any case, I was facing a user barrier, and I needed to overcome it if I was going to succeed.

The CPA who warmed to me first told me that other BAs had come in and tried to "fix things" only to cause upset and leave without ever really understanding the AS IS process, let alone providing TO BE solutions. That meant I had an even bigger hill to climb to gain the user trust I needed. But I wasn't a newbie when it comes to users. I knew what to do. I had to work hard to understand their domain, and my first task was that horrific file room.

For my first two weeks, I spent entire days in that room. I learned what documents were in a file and what they were for, and I learned the order in which documents were supposed to be filed. I learned the rules according to which the files were organized in the multitude of racks. As part of this effort, I also learned much of the vocabulary specific to these documents and the auditing domain. One of the CPAs had been appointed as my Subject Matter Expert (SME), and I took full advantage of her. She was agreeable and patiently answered my questions about what various terms and categories meant. As I learned more, I was able to ask useful questions, and this pleased her. She could tell how important it was to me to understand what the group did and how it worked. In particular, once I understood what was going on in the file room, we had long discussions about the problems with the paper filing system and the filing details that the automation would have to handle successfully, if the paper were ever replaced by soft files. And during these discussions, I could demonstrate that I actually understood what she was talking about.

With those two weeks of insight into the file room under my belt, I was ready to try talking to the individual CPAs again. This time, I carried a file from the file room with me, and I had some understanding of the documents in the file and the order in which they needed to be processed. I also had a million questions for my users, so I was definitely ready. Now, it was time for the second part of my effort to win them over. I had to demonstrate that I understood what they were saying and that, if I didn't, I would return after I did.

Most of my user meetings covered topics I could grasp, but some required me to go back to the file room and work with the SME to understand. The workflow steps involving the auditing of mortgage investments and municipal bond sales were particularly problematic for me. It was all very tricky financial stuff and the math was complex. But I stuck with it, and after a week or so, I won over my users. I was able to learn what I needed, and as with all users, my interest in the details of their work and my devotion to really understanding what they did created a strong mutual trust.

With trust came openness. Now, they talked to me and confided in me. They explained the pain they had with the current process, what they needed from automated software if they were to give up their documents, and the kind of security they demanded in order to have confidence in the accuracy of the automated process. They showed me the rudimentary software they were using in combination with the paper flow, and we discussed their issues with it. They were happy to identify their key needs and explain their most critical issues. They now looked forward to this project with hope that I could improve their daily workflow. Without any prompting, they reported back to their managers that I really understood what their process was, and they trusted that I would come up with the right ideas to help them. Halfway into the project, their management made a point of telling me how happy they were to have me there.

Six weeks the project, I understood their AS IS process and wrote up my report. I documented the current paper workflow, with examples and flowcharts, and described the points of failure. The project's Steering Committee reviewed the draft report several times, and I made changes as needed. So far, so good. They were happy with what I'd accomplished and wanted me to continue. I began the TO BE effort.

My instructions were to provide a TO BE analysis based on the best solution, without concern for any restrictions development might have. The Steering Committee would deal with such things later. This made me a bit nervous, but what the heck, there were IT people on the Steering Committee. If they were OK with it, I was happy to proceed. My TO BE report went through a number of informal reviews. As I came up with a section of TO BE automated workflow, I would ask the CPAs who would be involved with that section to review the flow. This review process slowed me down but was invaluable. I had to know if the CPAs who had the expertise relevant to a section thought that my flow made sense, that it could work, and that it met their needs. Often, these reviews required me to rework some of my ideas and flow designs, but it was all worthwhile. Once I completed the informal reviews for all of the workflows, I made final updates. Next, I designed a set of screens that implemented the workflows. I did this at a high level. I created screens for the major steps of each workflow. I didn't 't drill down to the lower level screens in each step.

The week I submitted the final report to the Steering Committee, a recruiter called me with a great job opportunity. I took the job, on the condition that I wouldn't start for two weeks. I wanted to see my current project through to the end. That week the Steering Committee met to approve or reject my report. The meeting included a few surprise attendees. Several developers from the state's IT department were seated at the conference table, in addition to the IT management that was always there. We went through the entire report, focusing mostly on the TO BE section. When we came to the pages with screen designs, there was pleased excitement around most of the table. Nothing helps business users understand a task flow better than a screen design that implements that flow. But the developers were not happy. They said they weren't expecting to see screen designs, and they couldn't commit to them. Further, they felt there would likely be problems with implementing a database that would support the features in some of the screens.

Typically, there was no real pushback on the developers' complaints. Even though the CPAs in attendance bravely reiterated that they liked the screen designs, there was no follow-up to their comments. I could see the discouragement in their eyes, and I knew what they were thinking. After all this, when we finally get a plan that will work, we can't have it. But they would have to accept their fate, as good bureaucrats always do. Without further ado, the committee chairman brought the meeting to a close, thanking me profusely for my excellent work. He recommended several future meetings to work through the details on the screen issue and any other questions that might come up about the report. In other words, his message was, "We aren't going to go against the IT department, so we have to figure out an alternative." Well, I wanted to give them all the finger, but decided instead to shake the dust off my feet and go on to the new job that was waiting for me.

About two years later the recruiter from that government job contacted me about another opportunity. I wasn't available, but was still glad to hear from him. We kibbitzed for a while, and he told me that my report for that previous AS IS—TO BE gig had been accepted with the screen designs, after all. He said the IT department had been told to implement my design, the software had been deployed, and the users were very happy with it. That was the last I heard, and even though I remain skeptical about this news, I choose to think positively and believe I helped those CPAs during that particular masquerade.

Government mask 2

Some years later, I took another state government contract as a User Acceptance Tester. That turned out to be my final masquerade. I was told I'd have control over the acceptance of the UI for a newly revamped government website. Most of the coding had been completed for the UI, and it was now working and ready for testing. There were two phases to my job. The first was to assess the usability of the new UI and report any issues that needed to be fixed. The second phase was to create the user acceptance tests for the new website, conduct the tests, analyze the results, and submit a report of my analysis. If I concluded that the UI had passed muster with the required sampling of users, the software would be deployed.

I took the job primarily because unemployment in Tech, and most everything else in the country, was high in 2011. Jobs were scarce. Even though UI Design had matured as a profession, no one was hiring for that role in my area. I had just been laid off from my dream UI design job—working with retail touchscreen software—and I was in the middle of a major home remodel. I needed income. This government job paid well and also gave me some chance of improving the UI, if it needed it. So, I slapped on the mask and joined the party.

The government IT group I worked with was typical of a bureaucracy. Everyone was in hunker-down mode, desperate to hold on to their jobs in a bad economy. No one wanted to rock the boat, cause friction, or say something that was critical or contentious. That included complaining about the new website UI. I was hired to do the complaining, if it needed to be done. Cool. They couldn't have found a better complainer than I. It was all good.

Before I joined, the IT group had convinced management to hire contractors to code the new website. The government employees insisted they were too busy to take on that additional work. Times were bad for taxpayers, but the state government IT coffers always seemed to have money available. So, an RFQ (request for quote) for the project was put out by the state. The developers they gave the contract to were a small company located in DC. Because the company was a veteran-owned business, several government contracts had already come its way, and it had a decent reputation. These developers, along with me and a PM, were the contractors on the project team. Several members of the state government IT group were also members of the team, but their roles were project oversight and liaison with management.

When I attended my first project meeting at the state government offices, the team had already been working together for a few months. As usual, the developers had called in from DC on the conference line. I could see that the PM was trying to work through the issues on the agenda, but he was constantly interrupted by the lead developer. This developer wanted to turn every coding issue into a non-issue. From what I could tell, his goal was to avoid additional work for his team because it would eat into his company's profits. His methodology was to insist that his team had tested their code, and it was working as designed. The PM tried to point out that the code that they recently delivered would not run on the state's test systems. The developer responded that this issue wasn't his problem. The state needed to check their setup because the error had to be there.

As the back and forth continued over the conference line, the developer would escalate his tone and volume. He had a deep, commanding voice, and when he eventually began shouting directives at the PM, no one had the courage or energy to make a peep. He would manage to shout down the PM on each issue attributed to his developers and then leave the call, declaring that he had another meeting obligation. He was a *prima donna* on steroids, and he was having his way. Then, he met me.

After that first meeting I knew, once again, that it was going to be war. Yes, I would try to be patient, calm and persuasive. I would try to coax him into realizing that fixing code to satisfy users was in his best interests. I would try diplomacy before war. But I knew that war was inevitable, because he believed he owned the UI turf and had the right to control it. That's who he was, and he would use his powerful personality to enforce his control. So far, it had never failed him. All the players eventually fell down before him. The funny thing was that I had met more challenging foes than he. This guy was gruff. He could huff and puff with the best of them. But he was not a great developer. He wasn't an innovator or an elegant code guy. He wasn't a true Lord. He cranked out code as fast as he could and hoped it worked. He was a grinder, and I wasn't intimidated in the least. My reaction to his yelling was: Meh.

As things went from bad to worse with the code setup, the manager of the state government IT group stepped in to force the developers to come over and get their code running. On pain of losing the contract, the belligerent lead developer (I'll call him Mark) finally did send his team to fix the code and get it working on the state's systems. Once the application was running, I started my usability analysis. This effort quickly became a basic code testing exercise. Many features on the screens did not work as specified in the requirements documentation, and some didn't work at all. I reported these early findings to the PM, who reported them to Mark a week before our next meeting with the full project team.

I was on the pre-meeting call and was ready for Mark's pushback. He started by denying the issues, implying that I didn't understand the requirements or how to use the application. I responded by sending him the PowerPoint document I was preparing for the upcoming project meeting. I suggested we go through it now so he could see the details of the issues I was going to report. The idea of dealing with error documentation rather than just verbal argument scared Mark. He agreed to look at my PowerPoint. We went through chapter and verse of what the requirements said the application will do and then reviewed my screen captures of the action sequence I followed to test the requirements. The screen sequences clearly showed where and how the application failed to meet the requirements.

Mark wanted to yell but what could he say? Yelling at words and pictures doesn't work. The data was right there for anyone to see, and Mark definitely did not want the project team to see it. At the same time, he didn't want his people to spend their time fixing these problems, some of which were significant. So, he tried to work me. He asked for time to make the fixes. He asked that I hold my report for a few weeks. I agreed to do so, but I told him I was going to continue my testing in the meantime. If I found more issues, I would document them and report them to the PM. We ended the call on that note.

As the next two weeks went by, Mark told us that his team was working on the issues I'd found and was going to deploy the fixes shortly. He did deploy new code, but it fixed only a few of the less important issues and left the main ones alone. I wasn't surprised, and I wasn't going to fall for it. This time, I gave Mark no advance warning. At the next project meeting I stated that I had found many usability issues with the application. Mark jumped in and said he had just recently deployed new code that fixed those issues, and I should review that code first before any further discussion. I replied that I had reviewed the new code and many significant issues had not been fixed.

He started yelling that I likely was not looking at the new code, but I was ready for that. The state government IT guy responsible for providing me with the new code was there, and he assured everyone that I had the new code. The state government team members asked me to send my report to everyone on the team, including the IT group manager. They then asked the PM to schedule a meeting specifically to review my report. Mark asked them to give him a few more weeks to review my report and try to fix any issues. I jumped in to say that this would seriously impact my schedule for user acceptance testing. Mark was furious at this, but I was not playing with him. He managed to wheedle one extra week, which I knew would not be enough time for him to fix much. In any case, our cards were now on the table. He knew I was a threat, and I knew I might be able to force him to fix the UI.

My detailed documentation of the UI issues was overwhelmingly convincing. The upper-level manager of the strategic government IT group made a point of thanking me for doing a great job and told me to continue doing what I was doing. To satisfy this manager while avoiding any work, Mark's group kept deploying new code that didn't fix much, while claiming it contained the fixes I'd asked for. We went back and forth this way for months. I held his feet to the fire, and little by little, his team eventually did make all the fixes I asked for. I boldly communicated the required changes through my own screen designs, which showed how the problem should be fixed. These designs were not a major rework of the overall design. I had come in too late for that. But given what I had to start with, I made sure that it worked as best as it could for users.

After many iterations, when I thought the UI was in reasonably good shape, I wrote the User Acceptance test scripts and conducted the tests. The results were good. The tests demonstrated that only minor changes would be needed to fix the few problems users had found. Mark wasn't arguing anymore, and our few remaining conference calls went smoothly. The PM had quit months ago. He told me he had never worked with such an impossible developer and didn't want to do it for one more day. He up and left for another contract job. So, it was just Mark and I on these calls. Hahahaha

He didn't even bother to ask for my explanatory error documentation anymore. Whatever I said was a problem, he agreed to fix. After 9 months of this project, we were all close to the finish line, with little fight left in us. The client was happy with the results, we would all get paid, and it was time to move on. The Tech job market was recovering, and before the final, minor issues were fixed and the state's new software was deployed, I left for a real UI Design job.

Some months later, a colleague from the state government IT group emailed me with the URL to the new website. I checked, and the software was up and running. It looked pretty good, and the last fixes I had asked for were done. All in all, though, that project experience left a bad taste in my mouth. "Why?" I asked myself. After all, I had successfully beaten down the chintzy, bullying developer, saved users from an initially dreadful experience, and done some good design work to improve user satisfaction.

The answer came to me quickly and dramatically. I'd just started a new job with the title of UI Designer and Usability Specialist. I was no longer wearing a mask, and I loved it! I loved having the clear mandate to design the UI. I loved being recognized by the project team as a professional UI Designer, whether they wanted me there or not. I didn't have to sneak UI Design work into the job anymore. I could openly claim ownership of my turf. This was the right way for me to be. The masquerade was not the authentic way to be. It had to be over. I needed it to be over. I deserved it to be over.

Fortunately, the time was right for my lonely masquerade to be over. As the year 2008 ended, the UI Designer role was finally settling into a well-defined profession. The role's professional title gradually morphed from UI Designer to Interaction Designer to, by 2011, User Experience Designer, or UX Designer. This role had teeth. As the job market improved, many companies wanted someone to fill this role. The need was real, and developers were getting nervous about it. The time had come for them to accept a UI designer into their camp and work together. It was not going to be easy for them. They were ready to be nice, to be accommodating, but they still weren't ready to relinquish the turf. They still wanted the final say over the UI.

With a couple of exceptions (the above story is one of them), from 2005 to 2014 I had a number of real UX Design jobs where I didn't masquerade. But I still had to fight for control of the turf. As I've described throughout this book, sometimes I won, sometimes I settled for less, and sometimes I walked out. One of my best wins was the UX Design job I took in 2007 with a company that made retail software that ran on a touchscreen. Major business touch screen applications were coming into their own. This Tech was cutting-edge, and I thought my touchscreen user interaction designs represented some real innovation. But by summer of 2008, just as I was patting myself on the back, Tech was throwing down some life-changing shock and awe.

No one was ready for it. Suddenly, we were all behind the curve. We were all left in the rear-view mirror. The Tech cars had skittered to a stop on the shoulder of the track, the drivers and passengers sat up in their seats and stared out the window, eyes popping and mouths open. What was that magical, fantastical beast gliding along like it owned the road? Was it a Tech car?

Oh yeah, baby. It showed the world what Tech was really about, and it showed me that the turf I wanted had always been mine. I came to understand, finally, that this precious turf would wither and die if I and other UI designers didn't start working the soil. We needed to take control and stop sniveling and whining that *they* wouldn't let us have it.

CH 8. The Touchables

Between 2007 and 2009, Tech gave people a new way to be in the world. Using this new Tech, people got answers to their questions about the world right away, no matter where they were. This Tech made it easier than ever to contact family, friends, lovers, doctors, lawyers, and drug dealers. No matter where they were, people easily found stores, gas stations, restaurants, and car repairs guys, right away. People worked and conducted business, no matter where they were. You told people right away if you were going to be late and exactly where you were on the map. If people needed help, they used this new Tech to locate the closest available help right away, no matter where they were. People knew what was happening in the world right away. If it was something bad, they used the new Tech to warn loved ones and friends about danger and risk early enough to save them. If it was good news, they shared the moment before it slipped away. In a flash people traded schedules and arranged meetings with old friends before they caught a plane and were gone forever.

The world had never been this kind to people. But there was much more. With this new Tech people never had to be alone anymore. People never had to be bored anymore. People never had to waste time anymore. It was easy to do something worthwhile, learn something interesting, take pleasure in a game, shop for birthday presents, explore new friendships, no matter where you were. People worried less about danger and risk; people had more fun; the world suddenly had more to please them; time passed more agreeably. It was a new way of living. There was a lightness in being human because the world looked better. It was better, and we were better within it—we had iPhones.

Where had this Tech miracle come from? Well, it didn't suddenly drop from the heavens. It was Tech, and so it was made by hardware and software engineers. The awesome difference with this Tech, though, was that before the engineers made it, they were given a UI design, and after that, they implemented the design they were given. Period.

The now famous Tech designer, Sir Jonathan Paul "Jony" Ive[26], with input from Steve Jobs, created a blueprint of the iPhone that showed, in pictures and words, how the device would look, the functions it would have, and how users would interact with it. The design was verified with the Apple engineers to determine if they could create the amazing Tech that would be needed for implementing this device. It was not impossible Tech to create, but making it work right was not going to be easy. That phone was going to need some dynamite elegant code and innovative thinking.

There were struggles, and through them all the design team worked alongside the engineers. Problems had to be solved, changes and trade-offs had to be made, but preserving the design remained primary to everyone. The first iPhone may not have been as fully-featured as the original design, but the heart and soul of the design were there. The features that would make users love the iPhone were there. No corners were cut on quality, no inferior code was permitted just to make a schedule, no complaints were made about the degree of innovation and customization required to achieve the design. No engineer decided to change the design, and no turf wars were waged over who controlled the requirements. There were no decisions by S&M to encumber the device with obstructive eye candy or useless features, or release the device before everything worked right. The design was given to the engineers, and they implemented it to the satisfaction of the designers. The code served the design. When the device was ready, it was released. If it succeeded, they would all take a bow. If it failed, the designers would have to take the heat.

The iPhone did succeed, for three reasons. Its design pleased users, that design was faithfully implemented, and the device worked right the first time. Those three reasons were powerful enough to cause shock and awe throughout the Tech world, leaving techies dumbfounded. How did they do that? How did they think of that? How did they know that would be so successful?

Hmmm. It's not rocket science. Understand users, design what they need, implement the design faithfully, and don't release until the product works right. Yet the first competitor to the iPhone—Google's Android—still didn't get it. The first Android phones were flops. The user interaction was a nightmare across virtually all of the Android implementations. People found them hard to set up and navigate. Compared to the iPhone, which had no learning curve and was easy and delightful to use right out of its lovely box, the Android was all wrong. It took several releases for Android phones to get it right, and that only happened when they started mimicking the user interaction of the iPhone—as much as they could legally get away with. The iPhone got its target users right the first time, and it became the gold standard. Users had spoken.

With the invention of gestures and new types of touchable selection widgets, the iPhone brought tremendous innovation to touchscreen interactions. Across the world people were excited about the new way of living offered by their iPhone and couldn't wait for the next iPhone release. Still, the business world was slow to jump on board. Many software companies viewed the iPhone as an entertainment and leisure device rather than a device capable of supporting robust, reliable, and secure business applications. Even several years after the iPhone's release it was hard for a software company to sell their business customers on the idea of switching from a computer with a monitor, keyboard, and mouse, to a touchscreen environment. I worked in the thick of this business mindset when I became a UI Designer for a retail software company. That job was a game changer for me.

8.1 Touch me, babe!

When I started that retail UI Design job in late 2007, the Tech world had recovered from the Bubble, but things weren't the same as before the Bubble happened. As I've discussed, most software companies were doing as much outsourcing as they could. It was rare for me to encounter the great developers of pre-Bubble days—those inventive creators of elegant code that drove innovation. Where were those Lords? After the Bubble, that herd had been severely culled. Many found careers outside of Tech. Some went into government work. Others could be found working as game developers in startups or hiding in place in small divisional offices of large companies, where they safeguarded the code that still worked. These "guardian" developers were typically kept around to monitor development schedules and fix the code that was delivered by the outsourcing companies. It was a sad situation, and, as annoying as they had sometimes been, I missed those Lords.

So it was that at my new design job, I was introduced to a skeleton crew of developers who were devoted to maintaining and updating older generation retail software. They kept it running, and the product had a small but loyal customer base. Our office was a minor division of a very large, global Tech company. As long as our division was somewhat profitable, we were ignored. A few of our developers and all of the managers were older, and they kept their jobs because no one else knew half as much as they did about the company's legacy retail software. There was a lot to know, too. The check-out desk software, the back-office software, the store enterprise software, the inventory and pricing databases, and the integration and communication software between and among all of them. But that was just the software side. The hardware side was equally complex, as you dear readers can imagine. Every suggested change or redesign to either hardware or software had to be evaluated in terms of its effect on all the parts of the system. The Tech was challenging, and I loved learning about it.

I also loved my managers. They were fully behind me, introducing me to the team as the UI Designer, who would be responsible for providing the screen designs to developers. On top of that, I would be working with them on a new and exciting project. The retail division had decided the time was ripe for a complete redesign of the check-out desk and back-office software, including the requirement that it would run on a touchscreen and be completely independent of a mouse and keyboard. The check-out person would be able to use the touchscreen to perform every retail function that was formerly available with the old software. I was also told that the higher-ups in our global company had a strong, vested interest in this project because touchscreen hardware was one of the company's major products. So, if we came up with amazing software that retail customers wanted, there was a good possibility they could also be talked into buying our hardware.

Scoring a real UI design job, learning about touchscreen Tech, working on the exciting new software project, and the top-level salary were all compelling feathers in my cap. I had stars in my eyes. On the other hand, the reaction of the 4 company developers to my role on the team was a whole different story. They were perfunctorily welcoming but took no interest in what I would be doing on the team. I knew they viewed me as a likely annoyance and, worst case, as a potential roadblock. I wasn't deterred, though. I had been given the turf I wanted, at least in name, and I would do what it took to win them over so they would cede that turf to me for real. No masquerading this time.

Things went very, very well. The new project was just beginning. One of my first and most important tasks was to get control of the outsourced graphics project, which had become a living hell for the manager of the development team. (I described this situation in the previous chapter, but I'll refresh it quickly here.) The manager had taken on the outsourcing task himself, thinking "it's just graphics, it will be a no-brainer, so I'll handle it instead of wasting any of my developers' time." But after 3 months of late-evening phone calls with bad connections, trying to communicate through a severe barrier of inarticulate English, and still getting no working graphics, he felt he was herding cats. At my initial phone interview, he seized upon my outsourcing experience and graphics knowledge. He described the problem he was having and asked me how I would handle it. I told him, and the next day I was asked back for a final interview that ended with a job offer. The development manager (I'll call him Greg) later told me he was thinking, "If all she does is fix this miserable graphics problem for me, it will be worth hiring her!" Greg's pain was my gain.

The second week at my new job, the team decided to change to an Agile development process. I had experience with it from my previous bank and pharma projects, and the developers were soon impressed with how smoothly I worked alongside them in our new, close quarters. Everyone could hear everyone else's conversations, so they all heard me on the phone with the outsourcing team, managing the graphics development. With me at the helm, it wasn't long before we were getting regular graphics deliverables in the right format, per the schedule that I put the outsourcing team on. This schedule included regular phone reviews with them, typically held at 8am my time, to ensure that they made the corrections as needed and updated their deliverables.

All of this back and forth could be heard by the developers, who tended to come in early. When the final graphics were delivered, they were on time, and they worked and conformed to the design. The developers were happy about this. Now, there were no more problems with putting the graphics into their code structure. And if a problem did come up, they could pass it off to me with confidence, knowing that I would get it fixed right.

As for the UI design of the screens, unfortunately, I came on to the project after a conceptual framework for the screen design had been approved with the outsourcing company. Money had been spent on that initial design phase, and there would be no major rework. Not an ideal situation for me, yet it was wasn't the worst design, and it had some strong advantages. Also, I had the freedom to move and resize some screen elements, and I controlled the screen text and icon design. In any case, since there was no time or money for major redesign of the conceptual framework, I had to work within those limits. At least I had strong management support to meet with users going forward, and this would help me figure out how best to satisfy their needs using the design framework we had. There were many more screens that had to be designed, based on the features available from the main screen of the conceptual framework. I had my work cut out for me.

As I studied the nature of the retail domain, I quickly learned that touchscreen applications were gaining hold in restaurants, but that was about it. Self-checkout touchscreen machines were just starting to be introduced in a few grocery and big box hardware stores, but they were slow to be accepted by store customers. They had a number of usability issues, based on the limitations of their first-generation hardware and software. A more difficult problem for our division's sales team was convincing retail executives, who were our prospective customers, to move to touchscreens for their standard sales check-out software.

The stumbling blocks were significant. Touchscreen hardware was costly compared with the computer and monitor systems these customers were currently using. In addition, all check out applications are complex and require training time for employees. Moving to a new touchscreen hardware setup, plus a new application user interface, would be a double whammy for training costs. Finally, in the non-grocery market, the sales staff had less turnover and often relied upon the experience of older managers who understood the business. Unfortunately, those folks dreaded major changes to the checkout systems. Many of them had experienced the move in the 80's from manual price entry on cash registers to the new Tech—digital scanning checkout with credit cards and the use of a sales desk computer / monitor / keyboard / mouse setup. They were young then and open to change, but even so, the changeover had been a nightmare. They made it clear to their store management that they were not excited about moving to the new Tech of touchscreens.

Grocery stores, on the other hand, typically had a younger crew doing checkout who would adapt well to touchscreens, but the cost of new hardware was much greater, given the number of checkout lanes per store in grocery chains. Finally, the retail business, more than any other industry, survives on extremely tight sales margins. We're talking fractions of pennies. To protect their fragile profit margins, they focus on containing operational costs. As a result, they avoid the large outlays of capital for major operational changes as long as they can. They are definitely not the kind of customers who are willing to take a chance on new technology. They will move to it only after it has proved itself with a few competitors. New Tech has to come with guaranteed operational savings, and no one wants to be the first to try it out.

Yes, touchscreen point of sale (POS) software was a very hard sell in 2008 and long after that, as well. Grocery store chains didn't start using it until much later. The major chain grocery store in my neighborhood finally bit the bullet in 2015, buying a cheap, ugly, hard-to-use touchscreen POS running on low-end touchscreen hardware. Sales personnel and customers alike suffered through that changeover.

Despite the difficult marketing obstacles, we did get a few customers for our touchscreen POS. They were not grocery customers, though. Instead, they were medium-sized department store retailers who were using our old POS application. One of them was interested in a pilot program using our new touchscreen POS in one of their stores. When I was hired, this pilot program was in the planning stages, and the customer was signing a deal for deployment of our first release in their pilot store. They made this deal conditional on a testing period, during which they would do dry runs of sales on real items in their store inventory.

For this testing, they would set up our first release in a back room of the pilot store, and no real customer sales would take place. If that testing was acceptable, they would move the system into one sales desk in the pilot store for use with customers. They would then run the pilot project for an agreed time period and, after that, they would make a decision about putting our new system in their other stores. At my job interview, I had emphasized the importance of face-to-face meetings with users, and the interviewing managers had agreed with me. To my delight, I was put in charge of scheduling visits at our office with users from the pilot store, who were anxious to learn about the new system. I was to meet with them to understand their key needs and give them previews of the new release to get their feedback. I couldn't wait!

In the meantime, our first release was progressing on schedule, and the developers were now relying on me to manage UI changes as issues came up. They had seen how I worked through UI changes and adjustments with the graphics deliverables. They liked what I did, so they were open to letting me solve other UI issues as they came up. I would explain why my solution to an issue would be better for users. They would be skeptical until they made the change and ran the software. Then they could see how my solution worked better than the old version.

After a few of these experiences, they started to trust me, and they were actually relieved that they could reliably shift the burden of UI design to someone else. I knew that they wanted the software to be successful. They wanted users to like it. They wanted to make a good product, so it was natural for them to be cautious about giving up control of their turf. But they did. They gave me that control and, eventually, they were even OK ceding that piece of the turf to me. I became the owner of the UI design.

"Hey Eva," one of them would say, "can you come over here and take a look at this? A problem came up, and I need to know what I should do on the UI." I would walk over, and he or she would show me. Then, after mulling it over for a few minutes or, for the complicated cases, maybe a whole day, I would say, "Here's what I'd like to do." I would bring a paper design of my idea and explain why I thought it was best, and the developer would just say "OK. I'll do that. I'll let you know when it's ready so you can check it out. Thanks." And that was that.

But things got even better as we added new features in subsequent releases. We all agreed that I would first come up with the new screen designs for using the feature, and then schedule a team meeting where I would explain how the screens implemented the new feature. We would discuss any potential code issues or questions about the design logic, I would make any changes we agreed to, and then the developers would write the code to implement the design. This process became our regular workflow, and the team looked forward to it.

Management would often attend one of the review sessions and would later express their appreciation for how this process was focusing the team on the users' needs. I and Greg, the manager of the project team, reported to the same director, although Greg was at a higher level in the company than I. The situation was good among the three of us. Greg and I had an open and honest relationship, and his respect for what I did and how I did it was clear to everyone on the team. I liked to solve problems myself, but a few times I had to go to him for help. I would say, "Hey, I need help," and his response was always just one word, "Anything." He had a great sense of humor too, which we all appreciated, especially during the tough times of the project. At first, though, when I was still a bit skeptical of everyone, he put one over on me big time. Here's the way that happened.

Whenever my previous jobs included responsibility for the UI Design, I would report design/usability issues as high severity bugs during the software testing phase. Developers, in my experience, didn't like that. When I first started reporting bugs to the retail project team, the developers showed their usual stank face. They were glad I was managing UI issues, but they still didn't think those issues were very important. One day I mentioned to Greg, in front of the team, that I had entered several high severity usability bugs, which needed to be addressed. With a straight face and a growling voice, Greg delivered one of his more famous lines, "The software was hard to make; it should be hard to use."

The first time I heard this joke, before I knew him well, I thought he meant it. After all, I'd heard worse than that from the Lords. I looked at him with despair. Haha, Greg was quite pleased with my reaction. He could see I had stepped right into it! He gave me a loud laugh so I wouldn't faint. As I was recovering, Greg turned to the stank-faced developers. He told them to fix the usability bugs, firmly reminding them that the success of our product depended on getting that part right. The issue of my bug entries never came up again. As the project progressed, I would often think back to that joke. Sure, it was funny for being so ridiculous, but that joke taught me a valuable lesson. It identifies a relationship between software and users which, when inverted, captures an important truth. Here's what I mean by that.

The touchscreen work our team was doing was cutting edge for business applications. The developers were using the new Microsoft development environment, which contained the latest and greatest programming power. They took full advantage of these tools and coded for pin-point performance and security. Touchscreen interaction has to be perfect or users will immediately become confused and frustrated. While users may be willing to click on a button a few times with a mouse to get a response, they refuse to keep poking a button with their finger. They expect the touchscreen to respond immediately.

Our developers also had to buttonize interactions that were previously done with a mouse, such as scrolling through a list and clicking to select items. The iPhone gestures were yet to come, so we had to design screen interactions and screen elements to work through button actions. As we progressed through several releases, both the developers who stayed with the company and also the few new hires that were made became better and better at coding these interactions. With a few releases under our belts, it started to feel like the old days. Our developers were extremely sharp and knowledgeable. They turned around changes quickly and reliably. Their planning and flow-charting capabilities for complex POS flows (like Return and Loyalty functions) were impressive. They fussed with each other over coding options and argued for elegant solutions. I saw them emerge from their second-rate roles as guardian developers of legacy code to become first-rate experts in touchscreen application development. Their work had demanded it.

As I reflected on how far we'd come, I would remember Greg's joke. In fact, he was right that our application was hard to make, and as we added more features for users, the coding difficulty increased. I realized that the easier we made it for users to perform their tasks by touch, the harder we made it for the developers to code. There was definitely an inverse relationship there, and I vowed never to forget it. In fact, I carried this idea forward in subsequent job interviews, enunciating it as a principle and an alert: The easier an application is to use, the harder it is to code. The reason this is true, I would explain, is that the code must contain the intelligence that frees the user from having to study how the application works. The application must make the user's tasks flow as the user expects. To do this, the application must anticipate each action the user wants to take and make that action the most obvious choice on the screen. That intelligence is hard to code, even when a UI designer provides a detailed screen design for it. Usability and delightful user experiences are critical for success, but they don't come easy. I knew it was important for management to know what they were getting into by hiring me.

Over the first two years of that retail UI Designer job I met with many users and designed many screens. I was learning more about touch UIs as I went along. Self-service, in-store kiosks were all the rage, and I created a prototype of a kiosk loyalty application for a prospective customer. I put the HTML prototype on a CD, and the sales guy and I traveled to the customer site. One of the kiosks that our company was trying to sell to the customer was already set up there in a large demo area, so I copied the prototype onto its hard drive, and we started running it.

My prototype demo had cool eye candy that was grabbing the attention of the execs as they strolled into the demo room. Once they stopped to look, it was very easy for them to start using the demo, and they couldn't resist. One of the upper level execs really loved it and asked if they could have it. I explained it was just a design demo, not a real application. He argued with me, "But it's working. I just saw it work!" I smiled like the Cheshire Cat. I'd done my part to sell the kiosk.

In addition to the kiosk demo, I came up with an innovation for POS training that excited the whole team and interested our customers. Our sales guys loved it. Even though our touchscreen POS was simple to use, our customers remained concerned about the learning curve for it and the costs of employee training. My innovation built the training right into the UI. With a simple button touch, a sidebar appeared on the screen that explained the operation of the screen currently in view. This functionality allowed the user to read how to use the screen while viewing that screen and interacting with it. It was a self-training system to be used for training as well as a dynamic help system accessible during an actual sale.

The developers loved the idea, and we worked together, thinking through the capabilities they could code into it. They came up with some great ideas that I didn't even know were possible, and we were all excited when they started the implementation. I supplied the design of the sidebar and the information that would appear in it. The finished product was a delight. Users loved it. The retail executives thought it was a great idea. I even published a paper about it in a popular UI Design journal. We demoed it at the annual New York retail federation conference, receiving glowing reactions from visitors to our booth. I was at the top of my game.

But soon after that high point, the Great Recession began to dismantle the retail industry, even as the iPhone showed us what a touchable UI should really be doing for people. By the end of my third year in the job, our division had failed to gain any significant new customers, and I realized I had a lot more to learn about designing touchscreen applications. There wasn't much new work coming in, and the future of my job looked bleak. It looked like my remaining time with the company was going to be short, with little to do. With things slowing down, I viewed the situation as an opportunity to investigate the UIs of phone apps, and I even began creating a prototype of a smart phone retail ordering app. But I was just messing around to see what I could learn from that exercise.

At the last quarterly meeting I attended in our office, the discouraged and sullen group of managers was discussing the bleak future of retail. Our Director said that we need to come up with new ideas to meet the changing retail customer. This statement was met with crickets around the room, so I decided to pipe up with an idea I had been kicking around. I said, "Why don't we create a grocery loyalty app for the iPhone?"

Everyone was startled and couldn't seem to get their heads around the idea. Finally, the one development manager known for his lack of productivity condescendingly declared that we aren't in the business of writing phone apps, and the functionality of a loyalty application can't be provided in a phone app anyway. Alrighty then. About 3 months after that, I was part of the first round of lay-offs.

A year later I was catching up with a former colleague at the company. She said that a few months after I left, the Director told this underperforming and condescending development manager to come up with something new for the company. The developer suggested a grocery loyalty iPhone app, which the Director endorsed. After almost a year he failed to deliver any working code, and the Director fired him. Yep, you just can't make this stuff up!

8.2 Metamorphoses

My in-depth design experience with retail touchscreen software was a wake-up call. The need for understanding users was now, more than ever before, glaringly apparent. Now we were dealing with direct human computer interaction. See, want, touch—the interaction had no intermediaries, such as a screen of text for the user to read, or a keyboard or mouse for the user to operate to access the software's experience. See the object, like the object, touch the object and Snap! you're rewarded with something you wanted to do, to see, to play with, or to buy. This touch modality of interaction was not only easy and direct for people, it was instinctive. They loved it, and I realized that touchscreen interaction had the potential for achieving great user satisfaction, if we provided the right UI designs.

Seeing, aiming, and touching a desired object is a highly developed skill that we master as infants, even before we learn language. The exceedingly refined evolution of human hand-eye coordination gave us the ability to throw spears and kill animals while running full out, delicately pick berries while avoiding their thorns, grab a can of tuna on the top shelf without tipping over the cans stacked precariously next to it, touch the shiny earring mommy is wearing on the first try, shoot a gun and hit a far-away moving target, and extrapolate that experience to rockets that go to the moon. I couldn't teach my 4-year old granddaughter how to use a mouse and keyboard to interact with a simple drawing application on my desktop computer. But she could use the drawing app on her mom's iPhone without any instruction.

So, yes, acquiring an object by touch is the first reaction humans have when they see an object they want. It's easy and enjoyable because it's an instinctive and gratifying movement for us. No wonder that when we use a well-designed touchscreen app we say, thank you, computer, you finally understand how I like to do things! Touchable software, I realized, was interacting with humans in the most fundamental way by appealing to their instincts—that is, their human nature.

Computer code, on the other hand, has its own nature, and it's not human. It's binary; it understands only restricted, unambiguous syntax; it does not love life or fear death. The chasm between the computer's nature and our own can be bridged successfully, but not by requiring humans to become more like computers. This truth was the most significant learning of my UI design career, and it was solidified in my mind through my experience with designing touchscreen applications and using an iPhone. That experience helped me complete my metamorphosis and take flight as a UI designer.

As I ascended, I looked back at the starting point in my UI design career. That starting point was the realization that users don't like Tech for Tech's sake. Instead, they like Tech only when it makes it easier and faster to do something *they* want or need to do. Once I understood this, my goal became figuring out how to make user interfaces that provide easier and faster experiences. I got in on the ground floor of this new way of thinking about users and user interfaces. It was fun to be part of a new wave, but the Tech industry's technophiles I had to work with were firmly in the opposite camp. They believed that it was the user's job to bridge the chasm between humans and computers. They thought it was the user's job to learn what input an application needed, and that it was the user's job to provide it with perfect accuracy.

In their view, the Cleanup Crew's role was to provide the training, support, and tedious documentation to users so they could learn the Tech needed to use the application. In my journey to the light, I told myself that coming to a project team as a UI Designer and user advocate immediately created battle lines, so I armed myself and was ready to fight. Hindsight is 20-20, but as I think back, I wonder if I was too quick to fight and too slow to try and persuade. In the end, I think that both have to be part of the monumental change that has begun. This is also a metamorphosis. Software teams are now starting a project by focusing on UI design. Their approach is to first design the human interactions that they believe will satisfy their target users. They are trying to learn about human nature—how the new software can meet their users on their terms. Only after completing that work will they develop the code that implements the UI design they come up with.

This new way of thinking about software ushers in an exciting time for Tech. It reinforces the extraordinary notion that the Tech race is endless and ever spiraling upward towards greater human fulfillment. In such an environment, a broad range of skills and talents is needed, especially those skills that help the project team understand their target users. I have described my own journey towards what I thought I could contribute, and how I tried to do it. Based on that experience, I know a similar path of self-discovery and redefinition of career goals awaits so many others who find excitement and challenge in Tech. I hope they go for it. There will be struggle but very little boredom. The iPhone gave us a peek into a Tech future that is, literally, AWESOME!

Speaking of Tech skills and talents, how will the Cleanup Crew fit into the new era of UI design first and code second? In this new era, the UX Designer controls the user interface design turf, but as I've tried to explain, good design is nothing without proper code implementation. Who will test that the implementation of the user interface is working as designed? Who will provide users with support when something goes wrong? Who will analyze and document user feedback for updates and new releases? Focusing on understanding users in the context of large-scale business applications isn't just about creating a great design. Testers, technical writers, customer support specialists, project managers, business analysts, and S&M staff can all play a role in ensuring the success of the design. They must also undergo a metamorphosis.

The UX Designer must stay involved through the life of the project, but he or she is only one person. One person can't manage all of the tasks required to support the design. Developers own the code, and they have always relied on the Cleanup Crew to prove the code works as intended. In the new era, the UX Designer will rely on the Cleanup Crew to prove the UI design is correctly implemented. The Crew will have to focus on both code *and* design, with the understanding that the design is sacred. The code may work, but if it changes the functionality of the design, that's a Priority One bug in this new era of Tech. With this change in focus, the Cleanup Crew becomes the guardian of the UI design turf as well as the code turf. Cleanup Crew members need to know their Tech, but now they will also need to understand their target users. They must learn the Tech the developers expect them to know, and they must learn the user information that the UX Designer expects them to know. In my past experience, as user interfaces increased in richness and functionality, the Cleanup Crew started to feel something was missing in their work. They began to notice when user interactions were confusing or hard to use, but they were rarely allowed to count these issues as real bugs because the code still worked fine. They felt the users' pain but could do nothing about it. In the future, as we strive for making computer interactions more human, the Cleanup Crew will need to be major user advocates.

8.3 The return to innocence

As the iPhone matured and touchscreen business applications found traction, we entered a new era of user interface design. But other dramatic changes also took hold in the Tech world, like cloud services, and applications that worked seamlessly across desktops, laptops, and touchscreen mobile devices. These applications (aka, apps) could appear on 3 different kinds of devices simultaneously. Moreover, if a user made changes in an app using 3 different devices, with the app running simultaneously on all 3, the app managed it without a whimper. To keep them secure and functional, these apps relied on spectacular innovation in the black arts of Tech: networks, communications, databases, and compilers.

With the cloud storing vast quantities of data, innovative developers started to tackle the age-old question of Tech: What do we do with all the data that businesses have captured over the years? It was time to milk that cow. Screamingly fast data analysis, data integration, data extraction, and inferencing from structured data were now all in play. Look there, Toto, robots, and microchip implants, and drones, Oh My! A new age of software development had begun. More than ever, software companies attracted, coveted, and rewarded great developers who could make this new Tech work for their customers. We are now seeing these great ones working again, and the veil of mediocre code is beginning to lift. We're confident that this new Tech is only in its infancy. The best is yet to come as this Tech blends with the new focus on what humans need from computers instead of what computers need from humans.

The old Tech expected users to interact with the non-human nature of the computer. Users needed to learn when and where an application needed input and the proper computer format for that input. But now, in this new era, the tables have turned. The new Tech views users as Tech innocents. The users of new Tech are now allowed to be innocent of any knowledge of the computer's nature. Users are allowed to be innocent of technobabble like "input," "program," "operating system," "bootstrap," "driver," "message code," "user guide," ".exe," "restart" and so many other non-human computer terms.

With its innovative power of touch, voice, and movement, the new Tech has the potential to meet humans on their terms. We can be innocent, as we were when we first looked at a computer, and that's OK. We don't have to know any technobabble to get what we need from a computer. With the new Tech, "usability" means providing user interfaces that truly understand our human nature, allowing us a return to technical innocence while being well-served by our computers. Whenever we experience that kind of usability, we won't ever want to be separated from our Tech.

CH 9. Designing for Human Nature

Alexa, Siri, and Cortana walk into a bar. Siri starts looking for her wallet. Cortana gets busy searching for cocktail coupons on the internet of things. Alexa finds a table for 3, orders their drinks, and starts blasting her electronica playlist through the bar's speakers. Right away, everyone wants to dance with her.

Only one got it right the first time.

Bringing touch and voice control to users is only the first step in designing for human nature. It's designing the features and task flows that users need that makes touch and voice control useful to people. Robotic dogs (in their current state) are fun for Tech demos, but useless for people. These robots can't (as yet) provide what humans perceive to be a dog's loyalty and its unconditional, subservient love. They can't sense when their master is afraid of someone and react by bristling, growling, and showing their fearsome teeth at the offender.

A robot dog can't tell when I'm feeling down and gaze lovingly into my eyes, while lying at my feet to comfort me. These machines may be robots, but they are not yet dogs.

What do we need from our robots of the future? Do we want them to look like familiar, organic life? Do we care more about making them look animal-like, or do we care more about making them useful? If a robot can converse with us, like the best-selling Alexa, it is certainly important for the voice to be human sounding. But more important is what the robot can do for us. We would not like Alexa's voice to go haywire and become unpleasant, but in the current Tech market, we'd still keep it around for its usefulness. Would we say the same about its competitors?

9.1 What humans want: A speculative exploration

We humans are complex, but we are learning more and more about ourselves. Like the other animals of earth, we have a nature and a set of behaviors that define us and are inescapable. One hallmark of our human nature is our stance toward the environment. We have the ability to adapt to the pressures of natural selection in our environment by eliminating those pressures rather than being passively changed by them. And the tool we use to eliminate environmental pressures is technology. We terraform the earth so we don't have to hunt for food. We air-condition and heat our living and working spaces (our nests) so we don't have to experience the temperatures of the native environment. We eliminate health threats so we can live next to one another at close quarters in vast, nested cities rather than in the larger physical separations that the natural environment would require for health.

We also use technology to satisfy the inescapable needs and desires of our hard-coded nature. As mammals and primates, we prize our individuality, independence, and privacy. We also create family nests within larger groups, and our personal bonding diminishes with remoteness from family and group affiliations. We often say to ourselves, "when the chips are down, family is what counts." Along with these bonding instincts comes fierce attachment to our personal and familial possessions and territory. We create inheritance mechanisms to preserve and extend our property and stuff through time. Grandma's wedding ring is precious to us in a way that no other ring can be.

As George Carlin once noted in a famous routine[27], "we think our shit is stuff, and everyone else's stuff is shit."

Our technology dotes on creating secure, personalized spaces in homes, apartments, condos, mansions, and estates. We can't get enough of protective technology that safeguards ourselves, our families, and our stuff. It's no surprise that the first driver of civilization was the need for protection. We realized that what we couldn't do for ourselves could be done if we formed a trusted community. But we used technology to go farther. Human nature needs society, but we barely trust family and friends with our stuff. We can't help but think that everyone may well be out to get us and our stuff.

But if the need for security is one side of the coin, the other side of our human nature is fun. More than the other primates, we need to laugh and smile with others, and we use technology to extend friendships and increase our options for entertainment. Sleep and rest are vital needs as well, which explains the significant focus of technology on sleep remedies and enhancements.

Apropos of this book is our use of technology to enhance the highly refined experiences we call "working." Humans need purpose, which was supplied in our early days by the ferocity of our struggle for survival. Hunting-and-gathering to stay alive in an unforgiving environment was a full-time job that required physical stamina, skills with tools, and planning. Because of technology, most humans are now relieved of that horrific job, so why are we still subject to discontent, boredom, and anxiety? Why don't we sit around like fat, contented zoo animals and let the abundant world feed and shelter us? These days, with all of our abundance, it's become hard to starve and live the short and brutish life of our ancestors. For most of us, you have to work at it. LOL Why dedicate our lives to developing complex careers? Why work if we can get away without it? What do we have to be anxious about?

The irony of civilization is that, having eliminated the pressure of a harsh environment driving us 24/7 just to stay alive, life just doesn't feel right. Without purposefully acting on behalf of our survival, we are fearful and unhappy. We need to feel in control of our survival, and work gives us that feeling. Because our human nature tells each of us that you can't trust anyone else with your safety and success, our instinct is to rely on ourselves to stay alive. In the end, each of us is driven by our nature to make ourselves number 1.

The psychological impact of this drive is that actively managing our personal survival is critical to feeling safe and optimistic about life. In the modern world where, for most of us, mere survival requires little to no effort, we use technology to help us create work. We've used technology to structure our world with business, which provides work for personal reward in a stratified economy. Business is the work of making stuff of value, creating services of value, and then buying and selling them. Work gives us purpose. It demands effort, skills, and commitment. In return it provides personal competition, team effort, and rewards that enhance our personal lives. Not exactly as fulfilling as bagging a woolly mammoth, but it satisfies the strong, instinctive need to control our personal survival. And work is certainly more pleasant, with more opportunity for success and longevity, and less stress and anxiety.

The inherent driver among businesses is competition, and technology is the best friend of competitors. In our time, we have Tech—the specialized form of technology that creates, presents, distributes, and services information. Whether you're making shoes or rockets or robotic hearts, the way you create, manage, and consume the information required to run your business can make or break you. Companies depend on their workers to use the Tech they buy to manage their sacred information successfully and increase productivity. In the end, in this age of Tech, the creation of value through work all falls on users. But if the Tech doesn't understand what a company's users need to do and why they need to do it, and if the Tech can't make it easier and faster for users to do their work, then the work of that company loses value. And it's Tech that has failed that company. The company's users haven't failed it.

In this new age of Tech that I've been extolling, we will expect our business Tech to be easy to use because it will be designed for our human nature. It may have several forms: passive and actively robotic. We will be able to touch the Tech, talk to it, and make it move around to help us perform our tasks. It will present itself as a touchable, listening, watching, video screen with useful robotic appendages—all designed to help us to work through and complete the tasks we care about. It will anticipate and understand the commands, touches, and movements we want to use to complete the steps in our task flows. This human nature-oriented Tech will complete those steps correctly and give us the results in the format most intelligible for *us*.

The more complex our tasks are, the more UX design effort will be required, and the more difficult the coding will be to implement the design. Human nature-oriented Tech offers opportunities for design innovation and elegant code that go beyond what we've already experienced. Unexpected human talents will be uncovered and new concepts will be invented to help Tech present a human-friendly interface. Yet, the basics of sound UX design methodology, as described in this book, will remain the same.

UX Design Basics

- Meet with users
- Understand what they need to do, why they need to do it, how they do it now, what's most important to them, what's working well for them, what is not, and why
- Capture that information in the right format to identify key needs that the new Tech must satisfy
- Write ideal scenarios for accomplishing those key needs, based on your understanding of how the users need to work
- Design the Tech experiences for those scenarios
- Implement the designs faithfully

9.2 *What we want most from Tech*

What humans want from Tech involves all levels of the human condition, from essential needs to nice-to-haves. Right now, in our time, we need Tech to help us solve the human problems that threaten the core of our lives. Ideally, the world of new Tech will look beyond its business interests and step up. But the solutions we need will only work if they are designed for our human nature. Hard-to-use can have no place here.

Tech, this is a call to arms! Bring on the cloud and the robots, the drones and the implanted chips, and the Black Arts of Tech. Give us the facial recognition, the retinal scans, the elegant code and your power of innovation. UX Designers, know your true craft and claim your turf. The human problems we face are a matter of life and death, of hope and freedom, of safety and happiness. Please, fire up the fastest, smartest, toughest Tech car you have, and put the pedal to the metal. There is no money we won't pay, there is no effort we won't support to fix these problems. But who knows, maybe you'll do a bunch of it for free because you care.

Help us protect our schools, save our kids

The drive to protect our offspring is firmly embedded in our human nature. We know the enormous advancements that technology has brought to the health and well-being of this most precious and sacred segment of humanity. Yet, in our current struggle with school safety, it's shocking how little we have called upon technology to help. We have thousands of office parks where, thanks to technology, hundreds of employees walk safely and securely inside and outside of one or more office buildings many times a day, all in a safe and secure environment. Why aren't we providing this same level of security for our children in their public schools? What wouldn't we pay for this level of safety or even more for our children?

The new Tech could help, but our political will is not there. The political arguments of our state and national legislators are never ending, and each faction wants its solution to be declared the right solution before they will do anything. In the meantime, we could drastically improve the safety of our children with technology, but we aren't doing it.

As of 2018 the US has roughly 92,000 public schools[28]. For the sake of easy math, let's call that 100,000 public schools. Let's say we spend an average of $1,000,000 per school to put in technology for safe access in and around the school grounds. That would cost us 100 billion dollars for the entire country. A drop in the bucket. We threw 700 billion at the TARP[29], 2.1 billion to set up the Obamacare website[30], and, in 2014 alone, we spent 3.6% of our GDP (17.43 trillion dollars[31]) on NATO, which came to 627.48 billion dollars.

Of course, we also spent billions to save GM. On a yearly basis we also spend billions on dubious academic research and foreign aid. Yeah, I think we can afford to spend *at least* 100 billion on our kids to set up reliable security systems in their schools. And that money is just to get started. We would need to pay a maintenance fee every year to keep the new security Tech updated and running smoothly at our schools. Yeah, I think the lives of our kids are worth it, don't you? But what could new Tech give us for our money? How safe could they make our schools?

It may not be possible or desirable to build high physical walls around every school, but can't we build invisible high walls if we use some combination of smart cameras, ground robots, and drones for patrolling the school perimeter? This kind of new Tech can provide an invisible wall with instant analysis and notification of any breach. Through this tech we can immediately send a warning or something stronger to any threat. Restricted access points at the school, perhaps at different distances from the school building, if needed, can be set up using smart tech instead of forbidding jailhouse structures. One access point could be placed at the parking lot/drop off entry point, and one at the building entrance. Students and staff can pass quickly through the Tech with instant recognition that uses facial systems, or badging and/or microchipped "smart backpacks". Metal detectors are used in office buildings, malls, and theaters. Why not use them at school entrances for the upper grades? Getting into school can be quicker than standing in line to scan a ticket to enter the subway. Durable microchip body patches with kid-friendly designs can be available for parents whose kids often forget their smart backpacks.

And what about these smart backpacks? The Tech should be able to know what's inside them, or better, what's not inside them and alarm immediately if something isn't right. Well, heck, while we're at it, why not make the lockers smart too? Of course, with this kind of Tech, the entry privileges of a suspended student could be removed instantly. That student would then set off an alarm immediately when crossing the school's outer perimeter.

Just as with our protected office buildings, the school's Tech would be able to tell us, at any moment, who is in the school and who is not. Exterior doors would unlock for entry only with the right tech identification. Of course, it's likely that a room with a security specialist devoted to monitoring the Tech would be needed. But it may turn out that such monitoring would be done at a remote location. The new Tech and minimal monitoring manpower would be more effective at detecting and stopping threats preemptively than 10, armed school resource officers stationed in the school. Unlike the new Tech, armed SROs are there as a final response to a threat that managed to enter the school and perform an unlawful act. Sure, it's fine to have them too, but with great Tech, they may not be required because the threat would almost certainly be stopped before gaining access to the students and teachers, and that would be a good thing.

Admittedly, managing the safety of large numbers of kids is not the same as managing adults entering office buildings. Kids get sick in line and puke all over. Kids have to go to the restroom NOW! Kids forget their badge, or their smart backpack, or their microchipped patch came off. Little kids start crying at odd moments, older kids start to scrap and fuss with each other. When these things happen, the kids can't just walk through the entry point. They have to be dealt with separately, which will make them late for class. Doesn't this mean that the schools will end up dealing with more problem entries than standard entries?

Maybe, but don't you think the new Tech can address that issue? I can think of ways and so can the reader. But I'm confident that the new Tech—the Tech that designs for human nature—can innovate something much better that what I can think of, something, even, that is beyond our wildest dreams. It would be great to know exactly when our kids have entered school and where they are at any time of the school day. Mom could have a conversation like the following:

> *Alexa, have the kids started class yet?*
>
> *Ms. K., Larry entered through the C2 access door at 7:43 am. Sharise entered through the C1 access door at 7:49. Larry is currently in hallway 6A and his locker is open. Sharise is currently in room 5 on hallway 3B and her teacher, Mr. Smith, is present.*
>
> *Alexa, did Sharise take her lunch into school?*

Ms. K., Sharise's lunch is in her backpack in her locker outside room 5 on hallway 3B.

In a safe, Tech-smart school, our kids can be kids again. They don't have to practice hiding from active shooters. They don't have to look at us and ask why we aren't doing something to stop the carnage, why we are letting them get hurt and killed, why we aren't protecting them like adults are supposed to. How sad and grotesque are we that we allow this horror to continue while we protect ourselves at work, but not them at school?

Maybe if our Tech heroes step up and offer their help during one of their frequent stops in Washington DC, our government will have the sense to accept the help and start fixing this frightening, dangerous environment that we force our kids to be in 5 days a week. It's not a local problem. It's a federal problem. It cuts to the core of our human nature, and it must be solved right the first time. The quality of the security Tech for our schools can be no less than the quality we demand for the TSA, the protection of our Congress, or the security of our nuclear missile sites. Go ahead and fight about solutions involving guns and mental health and armed teachers, but don't let your ideological agendas interfere with protecting our kids right now with the best Tech available. Failure to make schools safe must never be an option.

Help us make government records work

We humans form communities primarily for security, but once we feel safe, we enhance our communities with layers of social interactions and regulations. Not only do we need protection against external invaders of our communities, we also need protection against internal troublemakers. Our security now relies upon complex legal and military systems, which are governed by equally complex political systems. Our invention of business and its associated commerce has resulted in a complex monetary system, managed largely by equally complex banking systems. Our social interactions, daily work lives, and private home lives are all intertwined with these complex systems. They satisfy important human needs, and so we would expect Tech to be a major player in making them more and more useful to us.

The banking system has always been one of the first in line to ride the wave of Tech, from full computerization of accounting processes, to the inventions of credit cards and touchscreen ATM machines, to Wall Street's computer-driven day trading and money-making computer algorithms. Managing peoples' money is a lucrative and serious business. You can't make mistakes and emerge from them unscathed. When it comes to gathering data about peoples' money, tracking their transactions, and knowing the status of an individual's ability to pay for something on demand, the banking business is a Tech wonder.

Indeed, the banks continue to develop and use innovative Tech to provide not only instantaneous views into the status and movement of money, but also immediate responses to control its movement. All this requires gathering, storing, and processing unimaginable amounts of data with remarkable levels of security and secrecy. Yes, they know how to use Tech for big data, and they rely on Tech to keep them from making serious mistakes with peoples' money. By and large, they have earned our trust. Our government systems, on the other hand, have become a technological mess.

When it comes to handling data about its citizens, the government can't be counted on to get it right. Except for the IRS, which we can rely on to crunch all the data and collect our taxes no matter what, the government is failing badly to manage the information it has. One hand of the government often has no idea what the other hand is doing. The results are often tragic. For example:

- A teenager with a documented history of mental issues can practice shooting at a certified gun range.

- A woman on Social Security retirement benefits moves to another state and reports her change of address to the post office, but neither the Social Security Administration nor the IRS update her address in their databases. She has to contact each one to update it manually. It takes 10 business days for the updates to take effect. During that time a letter notifying her of critical, time-sensitive information is sent to her old address. The post office is not allowed to forward the letter, so it is returned and never reaches her.

- A man with a 15-year old felony conviction in one state purchases a handgun from someone at a gun show in a different state, then sells it to a friend, who is also a convicted felon. This felon robs a gas station and shoots the clerk.

- An illegal immigrant, who has been deported 3 times for violent behavior, re-enters the US illegally and kills someone.

- A legal immigrant with an outstanding court date is stopped for driving without a license and has no ID. His fingerprints are in the ICE database, but the ICE database and the fingerprint database used by the police are not integrated.

- A school reports a serious threat from a Facebook post to local police. The post was made by a parent of one of the school's students. The local police investigate, issue a warning to the parent, and make a record of the incident. This record is not available to the FBI database, so it cannot be flagged for immediate analysis and investigation. A few months later, the parent in question purchases a gun, goes to the school, and shoots a teacher.

- The case against a father accused of child neglect is dismissed in court. Two months later he is stopped on a Friday night in a neighboring state for speeding. The police see a pending complaint of child neglect on his record. The police can't verify his claim of dismissal until the court opens on Monday, so they hold him in jail over the weekend. On Monday morning he's found beaten in the jail's holding cell and must spend 2 days in a hospital. Unfortunately, his medical insurance will not cover the medical costs of temporary incarceration.

- A pilot certified to fly his own plane is given a DUI for drunk driving in his local town. The DUI goes on his state driving record, and the pilot is required to report the DUI to the FAA for review of his license. He doesn't report it. A week later he is drunk and flies with 2 of his children, crashing his plane into a house and killing 3 people. His DUI record was not sent to the FAA database so it could be flagged and the pilot's license immediately suspended, with immediate notification to the airport storing the pilot's plane.

If your credit card is suspended, how quickly is that information made available to banks, government agencies, restaurants, and retailers across the world? Why can't our government get its data act together like the banks have done?

While working on several government contracts and chatting with developers, I brought up some of these questions. There was a variety of responses. Some said it couldn't be done because government databases were too vast and disparate, with many of them structured using incompatible Tech. Others said government databases could be interconnected but the cost would be prohibitive and the results unreliable and hard to maintain. The lack of security and probability of data breaches was also mentioned, as was the fact that each government branch or organization is wary of sharing its turf with any other branch. The general sentiment was that they all have control issues, lack of funding, and inexpert technical leadership. Other concerns involved the end users, who some felt would not welcome the burden of learning to work with yet a new computer system for looking up or entering data. As for the general public, they are not in favor of a Big Brother government that has all their personal data available in a centralized system.

The responses that say the data problem is too difficult don't really hold water when we see what the banking system has been able to accomplish with our data. As for cost, it seems increasingly clear to me that the danger, financial damage, and unfairness to citizens caused by bad and missing data certainly outweigh the costs. Furthermore, the integration and interconnections of databases doesn't have to be done all at once and with all of our data. The Tech is available to accomplish the data revolution we need in government, while protecting privacy and securing databases. We don't want the usual government Tech disaster, like Obamacare websites and Medicare mismanagement. We deserve the same level of Tech quality and security that is demanded for our military weaponry and nuclear submarines.

The effects of government data failures threaten the safety of our communities as well as our freedom and independence as citizens in those communities. But new Tech that designs for human nature can stop the failures. We need to understand what the target users of an integrated data system need, how they work, their goals and major issues. We need to design database software and hardware according to the needs of different sets of target users—police, judges, court personnel, ICE, FBI, DMV, and all the others involved in our safety and legal protection. And we need to design these systems so they share data seamlessly to meet each other's needs.

If a Tech system is hard to use, it won't be used. We have to make it right the first time, or it will be ignored. How much voice control do police need to make their data searches and entries fast, accurate, and complete? How much touchscreen? How much robotic support? For example, the command, "Computer, do a background check on Driver's License 167254," may be issued by a policeman stopping a misbehaving driver or a clerk in a gun store checking an individual's background. A background check at a gun store should find all and only the federal or local data that can prohibit the individual from owning a gun. The gun store shouldn't see any other personal details of an individual, so the system must simply return a red flag and an indicator of the government source of each prohibition. The background check shouldn't take 24-36 hours, but it's reasonable for it to take 15-20 minutes in a gun store. The policeman, on the other hand, may need the check to take 5 minutes or less.

Liberty and justice demand that citizens always have free and easy access to their data records. The new Tech will understand this about our human nature and design for it—maybe something like this:

> *Alexa, am I qualified to purchase a gun in this state?*
>
> *Ms. K., please touch my thumbprint scanner.*
>
> *Thank you, your access to this information is verified. My search for this information will take 30 minutes.*
>
> *Ms. K, my search for the information you requested is complete. Please touch my thumbprint scanner to hear the results.*
>
> *Thank you, your access to this information is verified. Your record has one red flag pertaining to a local government regulation that prohibits gun ownership in this state. If you want me to read the details of this red flag, please touch my thumbprint scanner now.*

Help us make medical records work

Humans have strong survival instincts. We rely on science and technology to remove or incapacitate diseases that threaten us so we can continue our lives unscathed. In our time (and for the foreseeable future), the diagnosis and treatment of a person's diseases generates large amounts of data. Each disease and treatment event in a person's life becomes a datapoint that affects the success of future diagnoses and treatments for that person. Even within the best healthcare teams, diagnosis and treatment decisions are only as reliable as the data they are based on. If the data about a person's past diagnoses and treatments is incomplete or contains errors, future diagnoses may be faulty or treatment may be ineffective. People may not get well or they may die, despite their healthcare team's best efforts. In our time, the preservation and curation of our healthcare data has become a critical component of our health maintenance. Are we doing a good enough job managing this data?

Electronic health records[32] (EHRs) have been around since at least 2008. In their early days, the EHRs were used only by medical personnel, and patients had no access to them. Patients continued to be lost in the maze of their paper medical records. Every time you changed jobs, you moved to the health plan offered by your new employer. Often, the new plan was different from your previous plan. Often, the physician network available to you in the new plan forced you to change doctors. If that happened, you descended into medical records hell.

To ensure the continuity of your healthcare data, it was your responsibility to order a copy of your medical records from your previous doctors and verify that those records were sent to your new doctors. This process hardly ever worked correctly. The new doctors didn't receive all the records, or they didn't enter all your previous data into their system, or your previous doctors had only paper records and you had to pay them to make the copies and send them, or you had to collect the records yourself and deliver them.

Worse than that, even when you thought you had succeeded, you'd find out at your first appointment with a new doctor that he or she hadn't been notified of your new records so they hadn't reviewed them, or the new records hadn't been added to your file. Then you had to play the memory game of frustration and anxiety, "So, Mr. K., when did you get your last tetanus shot? Did you get the new pneumonia vaccine or the old one? Which antibiotic gave you the allergic reaction 7 years ago? Did the surgeon do a partial or complete resection? What was the exact value of your last PSA test?"

The situation improved somewhat when all patient data became digital, enabling easier and more complete transfer of patient medical records. Even so, what happened to that data when it reached its destination was unreliable. How well could it be integrated into the software used at the destination? There was no guarantee that doctors and hospitals all used the same medical record software. Significant failures of data transfer from one doctor to another still occurred on a regular basis.

To manage this problem better, patients started looking for more control over their own data. People thought that, if they could at least see all of their own medical records, then they could participate more effectively in their healthcare. If a new doctor didn't have your previous records for some reason, then you could reliably fill in the blanks because you had a copy of all of your records. People began putting digital copies of their medical records on their own computers. It was a start, but many of those records take up enormous amounts of disk space. With multiple CDs or DVRs filled with CT scans, x-rays, and MRIs, it was becoming impossible to manage your own record storage.

EHRs were introduced by a major health system in my area in 2011. They allowed patients to log in to their personal, secure web pages to see the data the health system had about their health, including notes from doctor visits and test results. Patients could also add data to their records by completing health questionnaires offered on their web pages. When a patient went to a doctor in this health system, the doctor saw the same information that the patient saw, but the doctor had a different user interface, so the information was presented differently. In addition, the doctor typically saw more detail than the patient was shown.

As you would expect, the early days of the system had glitches, but even so, it was a big step forward for patient health record maintenance. Unfortunately, since 2013, the lack of progress with systems like this across the country has become a serious problem. The user experience for patients has not improved much. The UIs are often confusing and inconsistent, as well as outdated. Stupid design elements appear that have no business being there. Developers are trying to make the patient record system work across all devices, but so far, it's a poor implementation.

Even more critical are the difficulties that doctors are having with their user experiences. They continually have problems finding specific online information they need about a patient during a patient visit. Doctors are under significant pressure to manage the time they spend with a patient, so they can't afford to devote that precious patient time to wrestling with hard to use search UIs. When software is hard to use, people don't use it. If doctors can't use their electronic health record software efficiently, they won't use it as it was intended, and they may not use it at all. This means that patient care is going suffer. We can't let that happen. We need to fix this mess.

At many of my medical appointments, my doctors and their nurses have problems locating specifics in my records while working at the computer, and they can't help but complain while asking for my patience. Since I love hearing from users, I will typically pursue the conversation a bit. They've mentioned more than once that the common response of their IT departments to their usability complaints is that more training and documentation will fix their problems. Yes, we've heard that before. That was the old way of things, when it was the user's job to learn how to provide the computer-babble input that the program demanded. But that time is over. Now we focus on what people need and how they want to work. The new Tech designs for human nature.

We need UX designers to meet with the primary users of medical records: patients and doctors/nurses. Following the right UX design method, these designers will learn the key needs of their users and create the ideal task scenarios for each type of user. After that, they will design the screens that implement those scenarios, and the developers will faithfully implement these designs. But that's not all.

We need to make the integration between the data of different corporate health systems reliable and user-focused. Either all health systems must use the same EHR software, or else different implementations of EHR software must be made compatible. Compatibility can't be just at the data integration level; it must also be at the presentation level. Users need to see their previous data seamlessly integrated into their current, well-designed screens. This presentation requirement demands a UX designer's control. The designer has to decide how to present imported data in alignment with the current UI structure while showing the user know that it comes from a different health system. The UX designer has to answer the critical questions: What do our users need to see on the screen? What do they expect to see? What makes sense to them?

As medicine advances, patient data increases in amount and in type. We live longer, and it's a primal drive of human nature to live as long as we can. Secure, reliable, and user-focused management of patient data is critical for providing the quality patient care we seek. It's also required for providing a high standard of care across all segments of our population. Any searching the doctor or healthcare professional wants to do for a patient must be easy and fast. Voice commands may well be appropriate in the private context of a treatment or consultation room. The system must anticipate the kinds of searches its users want to do and make them readily available. Outside of the healthcare visit, patients should be able to search their records easily by voice as well as touch. To make our EHRs work right, we need new Tech and UX designs for human nature.

Help us make cars into life-saving carbots

The mid-size American SUV today weighs at least 4,000 pounds[33]. The number of cars on the road continues to increase each year. The number of accidents also keeps increasing, especially since the use of cell phones by drivers. We drive 2-ton missiles around at high speed and stare at our phones instead of focusing on the road. Our distractions while driving include forgetting where we are and forgetting who's in the car with us. Our cars don't help much. They offer "infotainment" systems that are hard to use and tempt us to further distraction. We spend so much of our time in our cars, but they are basically unaware of our presence. They offer hands-free phone use, but don't require it. They make some beeps and flashes of light, but don't interrupt our loud music to warn infallibly about an approaching car in the lane we are trying to move into. They don't recognize when drivers go to sleep at the wheel or stop looking at the road for long periods. They ignore children who are strapped into car seats and left helpless in a hot car.

Instead of behaving like stupid, metal missiles, we need them to be our protective guardians while we are driving. It's not like car companies haven't been working on Tech-guided safety features. They have, but with the perspective of old Tech. First, they code the feature, then they try to interest humans in using it. Their new safety features are typically hard to understand, difficult to use, and not exactly what people need. The lane departure and lane centering feature, for example, detects when the car goes outside the lane and mildly steers it to stay within the lane, provided the lane markings are clear and the camera is not obscured by bad weather. Well, that's not going to do much good if the driver is asleep or texting on a curvy 2-lane road.

People need their cars to tell them when they are going outside the lane, but more important, they need the car to tell them when they are distracted and not focused on the road. No buttons should have to be pressed and no set up should be required to enable such a feature. Just tell me and make me pay attention! Self-driving cars may eventually become a reality, but until then, we need new Tech that is designed for human nature. This Tech needs to understand the features people need and design the interaction with those features based on how humans want to interact. Here are some ideas, based on my personal knowledge of human nature. See if they make sense to you, too.

Users of cars need to be able to talk to their cars to understand what the car's features are and what they can change or configure. Setup of features, for example, should be available by interactive voice control. Cars need to know when a driver's cell phone is present and put it under the car's control. For example, the car could be set up to allow only emergency calling while a user is driving and standard calling when the car is in Park. Cars need to know when a driver is not looking at the road for a significant period. Cars need to know if children are seated in car seats, where they are seated in the car, and if they are strapped in appropriately. Cars need to know if children are in a car seat when no one else is in the car. Warnings need to be by human voice with specific messaging instead of beeps of little lights hidden behind the steering wheel. Warnings need to override any other transmission coming from the car speakers. They should work whether an entertainment system is turned off or on and whether the volume is set high or muted.

Users' ability to turn off safety features must work in a protective mode. Cars must protect users from turning off safety features temporarily but then forgetting to turn them back on. Cars must protect users from turning off safety features when doing so presents high risk to their passengers—even though users may not understand the risk or care about it. For example, cars could check if a driver has a valid license before unlocking the startup mechanism. Parents of teens might want to go further and enable the car to test for impairment before startup. The car could still allow startup in a declared emergency, but the car could also report invalid driving to the relevant authority so appropriate action can be taken. Stopping illegal drivers before they hurt someone is what a robotic car would do if it is designed for human nature.

Navigation systems in cars have a long way to go. Their UIs are primitive and clunky, and don't allow configuration by voice. Many still require users to type information when setting up a desired route. The new Tech can take us beyond all that with interactive voice control between humans and a smart car's "navbot."

Driving your car can and should become a relaxing experience during which you rely on an intelligent monitor to enforce the necessary precautions that keep you and your passengers safe. The car's robotic presence should always be available to you through interactive voice communication. You should be able to count on it to watch your behavior and warn you clearly and effectively about driving actions you must take for safety. Instead of pointing our dumb, 2-ton missiles at other dumb, 2-ton missiles that are speeding along the road, hoping for good luck and competent drivers, we should be surrounded by an intelligent, early-warning safety shield. A car bot should be riding shotgun with us.

> *Carbot, do I have any messages from Jack?*
>
> *Ms. K, Jack left 3 messages for you. I sent him your standard 'driving now' response.*
>
> *Carbot, where is the nearest rest area?*
>
> *Ms. K, a rest area is on your desired route in 15 miles.*

With the new Tech that is designed for human nature, we could be comforted knowing that our children are safer in their car seats. For example, if we leave them alone in the car for any reason, the car bot could be set up to send us a text alert message and to get help if we don't respond quickly:

> ***WARNING!!*** *Children are seated alone in the car! The car doors are now unlocked! In 3 minutes, an alarm will sound!*
>
> ***WARNING!!*** *Children are seated alone in the car! The car doors are now unlocked! In 2 minutes, an alarm will sound!*
>
> ***WARNING!!*** *Children are seated alone in the car! The car doors are now unlocked! In 1 minute, an alarm will sound!*

and so on until those 3 minutes are up. Then, thanks to the configuration options we selected, the car bot would blare its horn and robocall 911. If we should ever make that rare, terrible mistake, forgetting just one time that the kids are still in the car with the windows up in 95-degree weather, miraculously, it could turn out OK. Our children would still be safe.

9.3 The never-ending story of Tech

Tech runs on a multi-lane racetrack. This book focuses on the lane of software applications for business users. That lane has been the focus of my professional consulting experience. And it's clear to me that business users are going to be involved in using the new software that will address the critical issues I've described in this chapter. Healthcare, for example, is a business, legal services are a business, and delivery drivers and taxis are a business. When business users need Tech, companies arise to sell Tech solutions to those businesses. The new Tech that we are moving toward needs developers, UX designers, and Clean Up Crews who bring new skills that focus on understanding how software solutions can tap into our human nature.

In this book, we've traveled through time. We've seen how Tech's business applications worked in the early days. When we looked back at the era of the giant mainframes, that old Tech seemed crude and disconnected from its users. But as we moved on in this book, we saw that even when the Tech became faster and more functional, the gap between users and computers remained frustratingly wide. With this in mind, I took you on my journey to UX design as the way to bridge that gap. I described some of my design failures and successes, and told you that they all convinced me that my journey had been the right way to go.

I couldn't have been more correct either, as we reviewed in this book what has happened in the last decade. That's when the world started to move into new Tech—touchable Tech, speaking Tech, and robotic Tech. That's when UX design showed itself to be the key to making computers interact with our human natures. Through good UX design, Tech started giving us what we need in the ways we expect and in the ways that delight us. I told you that the new Tech strives to hear just one thing from its users: This computer understands me!

If, as I assert, we are entering an era of Tech that understands people, the entire landscape of Tech development will change. The coming of the iPhone was just a taste of the kind of change I'm taking about. We will be able to form bonds with this new Tech, as we have with our smart phones, and piece by piece, this will change everything about how we live in the world. The Tech will become part of our personal space, allowing us to extend our personalities and project them in new ways through space and time.

As long as we can feel bonded to our Tech, we will always want more. We will want Tech to make our lives even more safe and secure, easier, longer, with greater freedom and independence, more interesting and compelling, richer in social interaction, and more fun. There is so much about our lives and the human condition that we still can't control, and that's where we will point our new Tech. From where we sit now in the human story, our needs seem infinite. Can we ever make all the Tech we need? The Tech raceway looks to be a never-ending road, wherever we exist and as long as we exist.

Just about everyone is now or will become a Tech user, but not all of us get excited about why and how Tech works "under the hood." For those who do, I hope this book has shown you that human-focused talents and skills are just as important to Tech as computer-focused ones. I've given you a look inside the process of business software development. I've also described the skills you need in order to participate as a UX Designer. And I've introduced you to the Cleanup Crew and the important roles they play in the software development process. If you want Tech to be your thing, I hope this book helps you jump into a car and join the race. Keep in mind that new Tech is always built on current Tech. If you are able to master an area of Tech that interests you now, not only will you have marketable skills for current jobs, you will also be in a great position to learn the new Tech when it comes along.

My story, as portrayed this book, is a perfect example of this. I started in 1980 by mastering the latest word processing Tech (Script). That knowledge prepared me for the next levels in that area of Tech—from BookMaster to HTML. From there, I could quickly master the tools and techniques used for the latest UI design Tech, such as the creation of high fidelity, interactive HTML prototypes. Even the biggest innovations have roots in the Tech that came before them. Alexa seems miraculous, but 2 foundations of its Tech—voice recognition (VR) and natural language processing (NLP)— started years before that amazing device came to be.

For all of us who are users, I've tried to explain why Tech has so often failed to get software right the first time, and why it's not the user's job to figure out how badly designed software works. Users have been made to feel that software is doing them a favor. They have been made to feel that their technical ignorance and disinterest in how computers work makes them unworthy to be users of the software that is given to them. Users are told that they fail because they don't care enough to learn the technical jargon that the computer needs from them. It's their fault that the software is hard to use. Users have been ignored by software developers and S&M people, who pander to the feature hype of customer executives and brush any mention of user issues aside with promises of training and documentation.

We've finally come to a time when users do have power, and I hope they continue to exercise it by endorsing the good software and exposing the bad. In our time, user reviews have serious effects on the success of software. Companies listen. Business users, in particular, have every reason and right to approach software from the perspective that it should be easy to use, and if it isn't, then the software is at fault.

I've tried to show in this book that there is a way to make software right the first time. And that means that there are no excuses for bad software. If someone ever tries to convince you to accept their hard-to-use software, just point to:

your smart phone, or
amazon.com, or
Google, or
the Uber app, or
Bank of America's desktop and mobile apps, or
Alexa, or

any other software that understands you and say, "They did it right, why can't you?"

AP 1. Career Building and Parenting

I begin this discussion with a few assumptions that I won't spend time defending. If the reader disagrees with these assumptions, then the rest of the ideas in this appendix are hardly worth exploring.

First, I assume that parents who meet our society's standards of loving, responsible caregivers are more likely to rear children successfully than parents who do not meet those societal standards. For the purposes of my discussion here, I will call those that meet society's standards "Standard Parents".

Second, I assume that the more that Standard Parents can be involved with their children's daily lives and serve as the primary caregivers during their children's formative years, the greater the chance that their children will become stable, self-sufficient, contributing members of society. I will refer to such members of society as "good citizens".

Given these assumptions, a strong case can be made for the value to society of promoting Standard Parenting. To make this case, we can start with the observation that our society is built on communities, and communities are built on the people who live in them. To preserve our society over time, our communities must be replenished with new adults, as the older ones pass on. It's self-evident that, to maintain what is good in our communities and make life better for the future, the new adults who join them must be good citizens of their communities. And where will these new adults come from? Obviously, a large number of them will be adults who grew up within the families of our society. Will they be the good citizens that we need? We know the answer to that depends, in large part, on the parenting they had as children

Per the assumptions I began with, it follows that the better job we do at Standard Parenting within our families, the better chance we have of providing good citizens for our communities. And we agree that, the better our communities are, the better our society will be. Thus, the case is made that promoting Standard Parenting is of serious value to our society.

Moreover, it follows from my assumptions that one way of promoting Standard Parenting is increasing the time these parents spend with their children. This will increase the likelihood of creating the good citizens we need in our communities, which in turn, increases the value created for our society.

Now let's explore these conclusions further and get to the heart of the matter. I've established that it's of great value to our society to make good citizens. I've stated that Standard Parents (as I've defined them) are the primary source of good citizens. And I've emphasized the obvious conclusion that increasing the time spent in Standard Parenting will increase the likelihood of creating good citizens. But this is where the dilemma arises: Must we choose between promoting Standard Parenting and enabling parents to build successful careers? How can we maximize time for Standard Parenting while employing those parents in careers with high demands on their time? Is it OK to settle for providing only a daycare option for parents, which requires both of them to be away from their young children all day, 5 or more days a week? Is that really the best we can do to promote Standard Parenting?

From my perspective, it seems clear that we should not settle for offering only daycare as a parenting option, especially if we want to encourage parents (particularly women) to build a time-intensive career. I maintain that it is in our best interest to provide other options. And some of this is being done now. Time-sharing of a single job, for example, is a step in the direction I'm going. But we can do more. Here are some ideas that I believe are feasible for office-based employees, especially in Tech. They may also make sense for other kinds of jobs. While I offer them as a solution for women and men who want to build professional careers, I think they can also apply to non-professional occupations.

Flexible Remote/In-Office option with Onsite Childcare

Ideally, this can be available for both parents, even if they have different employers. The work week is structured to have 1-3 days of the work week spent at home, working remotely. If both parents used this option, they could arrange the work week so that one parent is always at home. Many of today's companies already encourage remote work, especially within global corporations where time zones across sites cause incompatible office hours. The remainder of the work week would be spent in the office, but childcare for infants to preschoolers would be available onsite. Furthermore, child visitation periods would enable parents to set aside several, hour-long periods during the workday for undisturbed time with the children. This time would be available during both remote work and in-office work. During these periods, parents could feed small children, read them a story before naptime, attend to doctoring for sick ones, and if working at home, greet older children when they return from school. The workday would be elongated by a few hours, but many Standard Parents would still welcome this option. And just like any parent working in today's offices, if an emergency arises, the parent can attend to it immediately. Finally, during the visitation period at the office, parents can have their onsite young children with them instead of at the onsite childcare facility. If the office is a cube setup, there can be a separate space for parent one-on-one time. In a remote/in-office option, the stress on the parent at the office is greatly diminished, knowing that the children are either nearby or else in their own house with one parent available to them. It's a win-win for career building and Standard Parenting time.

Nanny Services option

Even if a parent is working at home, it doesn't mean they can be working and caring for their children simultaneously. Particularly for on-demand careers, a parent at home needs someone else there to attend to the children. It's likely that the services of a nanny would be cheaper than the cost of a daycare center, but there doesn't seem to be much employer support for nanny services. There are many benefits to nanny services, however. First, on days when a parent can work from home, the children have the comfort of their parent around, the parent has the comfort of being available for any serious issues, and the nanny can be instructed to follow the parent's daily plan for the children. What they eat, when they eat, when they must nap, when they go outside to play, where they can go out to play, what clothes they should be wearing, etc. When the children are in daycare, parents never know for sure how well their plan is being followed or whether their children are getting the right amount of attention. Furthermore, parents are limited in what treatment they can require for their children. In daycare, the caregivers can only do so much. The option I'm proposing would allow parents to contribute part of their earnings to a Nanny Services fund---ideally, their employer would also contribute. A reputable Nanny Services firm would be vetted by their employer, and the parents would be able to select a nanny from this known, qualified source. Monitoring equipment would be provided, so that when in the office, parents could observe and communicate with their nanny. I see this option as being especially valuable to single parents with on-demand jobs, allowing them to be around their young children when working at home and stressing less about their childcare when in the office.

Regular Time-At-Home option

Many on-demand jobs that require constant attention are also highly structured. When you are on the job, you are totally on the job. But when you are off the job, you are totally free. Both professional and non-professional occupations call fall into this category. I suggest that these types of jobs can allow a parent to have regular time-off at home. For example, a parent can be given 5 days in a row off, per month. This time would be for regular childcare at home. It would not be vacation time, and the parent would not use it away from home. It would not be the official Family Leave time, either. It would be time that the parent provides childcare at home. For example, if the parent normally works 52 weeks a year, this option would reduce that time to 38 weeks per year (with vacation and holidays reducing it further, of course). But this option would not reduce the parent's employee benefits, and it would not reduce the parent's job to part-time status. If two parent households could take advantage of this option, it would allow a parent to be home for 2 weeks of every month. Even while at home, the parent may well have some time over during a week to devote to career enhancement. There would likely be a reduction in salary for this option, but Standard Parents (as I defined them earlier) would weigh that salary drop against the value of being there for their children on a daily basis and very likely think it well worth it. In addition, they would have the comfort of being at home with their children instead of incurring the cost and other downsides of daycare.

The Ethical Employer

I've offered some ideas for maximizing parenting time, and these ideas are not mutually exclusive. They could be combined in various ways to provide more time for Standard Parents to devote to their children and families. The options available could be configured to meet the needs of parents with preschool age children and those who have only school age children. But the bottom line is that ideas like these will require active financial support, encouragement, and open-mindedness from employers.

I am not claiming that promoting parenting time is the only way to create good citizens. But I am claiming that both the evidence we have and also our common sense tell us that the more time parents have for quality parenting, the greater the likelihood that their children will become good citizens. Of course, how parents rear their children is their decision, and one size does not fit all. Some may decide that daycare is the best option for their family. The key is to provide a variety of options, especially options that allow parents to give more time to their families while still progressing their careers. Daycare should not be the expected or only solution. Options like those I've discussed will not eliminate all the stress of parenting, and they will require monetary sacrifices from both parents and employers. But the rewards to both would make up for the money. Employees with less stress perform better and are more loyal to an employer who supports their needs. That's worth a lot to employers. Can more women succeed as a bleeding-edge Tech developer while having a family to care for? Yes, if we provide more options for balancing the needs of both.

On the news or in social media, we often hear businesses declaring their values and their ethics, insisting that they care about their employees and about giving back to their communities. Supporting and implementing ideas like those I've described would go a long way toward showing that they care. In my view, all businesses should support a very important value of our society: the value of quality parenting. Give me a business that helps its working parents create good citizens---now that's an ethical employer.

AP 2. Business Analyst Job Description

I have redacted the names and locations in the following job description, which I received in email in 2019. Other than the redactions, the text is exactly as I received it. I have not corrected typos or spelling and grammar errors.

Job Opportunity : Business Analyst in [redacted]

[redacted] via bullhornmail.com
Wed, Jan 9, 2019, 12:16 PM
to me
Hi Eva,

Hope you are doing great!

Your resume recently came across my desk while searching for a Business Analyst in [redacted] . Our client is currently looking to hire a consultant, and I think you could be a great fit! I've attached a copy of the job description below for your persual.

Job Title: Senior Business Analyst
Job location: [redacted]
Job duration:12 Months of contract (possibility of extension)

Required Skills

- 10 years of experience gathering and documenting business requirements
- 5 years of experience in writing detailed Business Requirements and Functional Design documentation

- 5 years of experience working directly with senior stakeholders to determine improved business solutions
- 5 years of experience working with cross functional teams
- 5 years of experience creating process flow diagrams
- 2 years of experience using Visio to create BPMN diagrams

Desired Skill

- 2 years of experience of Election Systems Experience
- Experience serving as the Lead Business Analyst on combined team of analysts and developers
- Strong understanding of the CRM Platform and its functional modules
- Delivering and leading end user training sessions
- Knowledge of Agile methodologies (SCRUM)

RESPONSIBILITIES

- Creating documentation that reflects business expert input and is suitable for use by developers
- Gathering, refining and documenting detailed business requirements
- Documenting as is and to be business process flows
- Business Rules, Functional Requirement, Process Flow, Use Case, and User Story documentation
- Defining Acceptance Criteria for each requirement
- Ability to review Functional Design to ensure it meets requirements
- Leading and mentoring less experienced business analysts
- Facilitate and accelerate the interpretation of business stakeholder input into developer ready documentation

Regards,

[redacted] - Technical Recruiter

AP 3. Use Case Example

There are many use case examples available on the internet and in books. I have copied one here that is similar in structure to the use cases I worked on in my bank project.

This example is the copyrighted work of Tyner Blain. I copied the example directly from, and freely available on, the following website of Tyner Blain:

http://tynerblain.com/blog/2007/04/09/sample-use-case-example/[34]

The material I copied and put here is licensed for use on the Tyner Blain website as follows:

This work is licensed under a Creative Commons Attribution 4.0 International License[35]

I do not claim, by copying his material under this license, that Tyner Blain endorses or warranties any part of this book.

****Start of copied material****

Here's an example of a use case that has some system complexity. The user interacts with the main system that we are describing. The system also interacts with two external systems. This use case example shows how to describe the steps that demonstrate all interactions with the system.

Sample Use Case Example

The first step in defining a use case is to define the name, using the verb-noun naming convention.

Use Case Name: Place Order

The next step is to define the use case at a low level of detail. This quick use case definition allows for agile development of use cases. This is also known as a use case brief.

Sample Use Case Brief

Use Case Name: Place Order

Actors:

- Shopper
- Fulfillment System
- Billing System

Use Case Description: After the user has selected items to purchase and then order the items. The user will provide payment and shipping information. The system will respond with confirmation of the order and a tracking number that the user can use to check on order status in the future. The system will also provide the user with an estimated delivery date for the order, which will include all selected items. The user may already have an account with the company with billing and shipping information.

Formal Use Case Example

You can clarify the use case further by refining the use case into a formal use case or informal use case format (free template). Here is how the use case would look when using a formal use case format.

Note that there are additional elements to a formal use case that may be included, but are not part of this example. You can read more about how to read a formal use case. You can also read about how to define actors, and how to define use case triggers and preconditions.

Use Case Name: Place Order

Actors:

- Registered Shopper (Has an existing account, possibly with billing and shipping information)
- Non-registered Shopper (Does not have an existing account)
- Fulfillment System (processes orders for delivery to customers)
- Billing System (bills customers for orders that have been placed)

Triggers:

- The user indicates that she wants to purchase items that she has selected.

Preconditions:

- User has selected the items to be purchased.

Post-conditions:

- The order will be placed in the system.
- The user will have a tracking ID for the order.
- The user will know the estimated delivery date for the order.

Normal Flow:

1. The user will indicate that she wants to order the items that have already been selected.
2. The system will present the billing and shipping information that the user previously stored.
3. The user will confirm that the existing billing and shipping information should be used for this order.
4. The system will present the amount that the order will cost, including applicable taxes and shipping charges.
5. The user will confirm that the order information is accurate.
6. The system will provide the user with a tracking ID for the order.
7. The system will submit the order to the *fulfillment system* for evaluation.

8. The *fulfillment system* will provide the system with an estimated delivery date.
9. The system will present the estimated delivery date to the user.
10. The user will indicate that the order should be placed.
11. The system will request that the *billing system* should charge the user for the order.
12. The billing system will confirm that the charge has been placed for the order.
13. The system will submit the order to the *fulfillment system* for processing.
14. The *fulfillment system* will confirm that the order is being processed.
15. The system will indicate to the user that the user has been charged for the order.
16. The system will indicate to the user that the order has been placed.
17. The user will exit the system.

Alternate Flows:

3A1: The user enters billing and shipping information for the order. The user desires to use shipping and billing information that differs from the information stored in her account. This alternate flow also applies if the user does not maintain billing and / or shipping information in their account, or if the user does not have an account.

1. The user will indicate that this order should use alternate billing or shipping information.
2. The user will enter billing and shipping information for this order.
3. The system will validate the billing and shipping information.
4. The use case continues

5A1: The user will discover an error in the billing or shipping information associated with their account, and will change it.

1. The user will indicate that the billing and shipping information is incorrect.
2. The user will edit the billing and shipping information associated with their account.

3. The system will validate the billing and shipping information.
4. The use case returns to step 2 and continues.

5A2: The user will discover an error in the billing or shipping information that is uniquely being used for this order, and will change it.

1. The user will indicate that the billing and shipping information is incorrect.
2. The user will edit the billing and shipping information for this order.
3. The use case returns to step 3A1 step 3.

10A1: The user will determine that the order is not acceptable (perhaps due to disatisfaction [sic] with the estimated delivery date) and will cancel the order.

1. The user will request that the order be cancelled.
2. The system will confirm that the order has been cancelled.
3. The use case ends.

Summary

This example use case shows how a use case can be developed with incrementally increasing detail. First, the name of the use case is identified, then a simple version of the use case is defined. Finally, a formal use case is defined.

****End of copied material****

AP 4. My Early UI Design Publications and Presentations (redacted)

2005	April, "From Technical Communicator To User Interface Designer," with [redacted] for [redacted] Conference, Society for Technical Communication, [redacted]
2004	December, "Managing Agile Development Projects that Satisfy Customers" for the [redacted] Project Management Institute – IS LIG, [redacted]
2004	(May-June) "Revisiting the Path to Usability" with [redacted] *IT Professional Magazine of the IEEE.*
2003	(November) Half-day Seminar: "User Profiling: The Starting Point for Delivering Information That Satisfies Your Customers" for the American Medical Informatics Association conference in [redacted]
2003	(October) "Getting It Right the First Time" for the [redacted] NAWBO Business Conference.
2003	(May)Full-day Seminar: "How To Design Software and Documentation That Customers Want To Use" with [redacted] for the STC Annual Conference [redacted]
2003	January: "How To Organize and Lead a Software User Interface Design Team" with [redacted] STC Telephone Seminar, sponsored by the Society for Technical Communication, US and Canadian audience in attendance: over 400.
2002	October: "Finding the Customer Behind the Veil of Requirements" for the [redacted] Software Process Improvement Network (SPIN) group [redacted]
2002	April: "Technical Communicators as User Interface Designers" for the [redacted] Conference, sponsored by the Society for Technical Communication [redacted]
2000	Leadership Award for my design work on the first release of the [redacted] product.

AP 5. Masquerading Job Descriptions

This appendix contains 3 redacted job descriptions I've received in the last year or so. You can see that the job requirements truly call for the skills of a UI Designer, even though the job title is masquerading as something else.

JOB 1. This job asks for a visual (graphics) designer to do the work of a UI Designer.

****

Visual Designer

job summary:

DIRECT HIRE FULL TIME

We are currently seeking quality individuals who can help us grow, and who desire to achieve a professional and rewarding career opportunity. We offer top industry salaries, and a competitive benefits package that includes medical, dental, vision, a 401k program with employer match, tuition reimbursement, and much more! You will always be working with the latest technologies in this role.

Please submit a portfolio outlining your Visual Design experience when submitting a resume.

location:
job type: Permanent
work hours: 9am to 5pm
education: Bachelors

responsibilities:

Ideal candidates are methodical explores of visual design and stay informed on trends. They inspire their teams with clean, grid-based visual design. They understand the role of research are exceptional communicators, and continue to actively develop their visual design knowledge/skills.

qualifications:

- Create a shared understanding of the solution among cross-functional colleagues to produce world-class design, advocating for the end user.

- Sketch the ideas to help everyone see and understand the concept; participate in regular collaborative sessions as an essential part of maintaining team rhythm, unity, and focus.

- Generate the appropriate visual assets from low fidelity to high fidelity deliverables based on project needs.

- Advocate for and participate in design research as needed

skills:

- BA/BS degree in Visual Communications, Graphic Design, or related field, and 3+ years of relevant experience; or MA/MS degree in Visual communications, Graphic Design, or related field, and 2+ years of relevant experience; or 5+ years of relevant experience. (Transcripts required

- Excellent written and visual communication in your portfolio; Confident verbal communication and presentation skills; strong sense of layout, visual hierarchy, color and typography.

- Familiarity with Agile methodologies, and a proven ability to work collaboratively with others.

- Strong understanding of screen-based design regarding web, mobile, and rich client applications.

- Experience building and or leveraging a design system (style guide)
- Experience with common tools for visual design and collaboration. (Good skills in Adobe Photoshop/Illustrator or Sketch, InVision, and Axure a plus)
- Please submit a portfolio when submitting a resume.

Equal Opportunity Employer: Race, Color, Religion, Sex, Sexual Orientation, Gender Identity, National Origin, Age, Genetic Information, Disability, Protected Veteran Status, or any other legally protected group status.

- location:
- date posted: Thursday, October 10, 2019
- job type: Permanent
- industry: Manufacturing
- reference:

****

JOB 2. This job asks for a Business Analyst to do the work of a UI Designer.

****

Req. ID: Sr Business Analyst
Position Closes: ASAP
Send Responses to:

Rate: DOE, Job Type: Contract, Start Date: ASAP, Duration: 6-12+ months, Number Of Openings: 1, Location:

Please Note: In-Person Interview REQUIRED on 1st Round – No Telephone or WebCam Allowed

Description:
Our client's Information Technology Division is seeking a Business Analyst to work with internal staff to optimize processes and improve quality for the design, development and maintenance of technical assets.

It is critical for a Business analyst to understand the operational systems of a client and what the client requires as changes to their existing systems. The Business analyst must translate those requirements into technical specifications to be used by internal staff to develop the system modifications; diagrams must also be prepared explaining the modifications so that both technical staff and executive management can understand what must be done. The Business analyst must propose alternative solutions that balance cost and benefit, and adequately test the possibilities for business and technical compatibility.

Responsibilities:

- Understand the operational systems of a client in order to discover and document required modifications in the form of User Stories to improve both business and technical processing

- Work with Product Owner to understand business requirements and be able to write them into Epics and derive required features from each Epic

- Create clear and concise User Stories for each Feature that are easy to understand and implement by the development team; Map stories to technical design specifications to be used by internal staff to develop the system modifications

- Identify and group related User Stories into themes, document the dependencies and associated business processes

- Assist Product Owner in maintaining the product backlog and progressive elaboration of features and User Stories

- Create conceptual prototypes and mock-ups

- Conduct Root Cause Analysis to identify system gaps and deficiencies that require a business, technical or blended approach for resolution

- Diagram proposed modifications using architecture and data flow diagrams, so that both technical staff and executive management can understand what must be done

- Propose alternative solutions that balance cost and benefit, and adequately test a solution's feasibility for business and technical compatibility

- Document current state processes where documentation is missing or out of date

- Collaborate with staff, vendors, consultants, and contractors as they are engaged on tasks to formulate, detail and test potential and implemented solutions.

- Perform Quality Analyst functions such as defining test objectives, designing test plans and test cases, and executing test cases (including regression testing)

- Assist Product Owners with testing the deliverables as needed

- Must be proficient in business process management concepts on BPM tools. Strong experience in creating business process models.

- Other tasks as assigned.

Qualifications

Required:

- The candidate must have a minimum of 10 years of experience delivering business and systems analysis artifacts.

- 5+ years experience as an Agile Business Analyst; strong understanding of Scrum concepts and methodology

- Experience organizing and maintaining Product and Sprint backlogs

- Experience translating client and product strategy requirements into application requirements and user stories

- Defining acceptance criteria and managing acceptance process with development

- Experience with large, multi-module systems

- Microsoft Office products (Word, Excel, Access, Outlook, Visio, PowerPoint, Project Server)

- Application Life cycle management (ALM)

- The candidate must have exceptional written and oral communications skills

Preferred:

- CBAP and/or CSPO Certification

- Business Process/Analytics

- Experience using Team Foundation Server for Agile Software Development and work item tracking

- Unified Modeling Language

Required Skills:

The candidate must have experience delivering business and systems analysis artifacts. - Required - 10 Years

Experience with enterprise, multi-module systems. - Required - 8 Years

The candidate must have exceptional written and oral communications skills - Required - 8 Years

Experience as an Agile Business Analyst; strong understanding of Scrum concepts and methodology - Required - 5 Years

Experience organizing and maintaining Product and Sprint backlogs - Required - 5 Years

Experience translating client and product strategy requirements into application requirements and user stories - Required - 5 Years

Microsoft Office products (Word, Excel, Access, Outlook, Visio, PowerPoint, Project Server) - Required - 10 Years

Application Life cycle management (ALM) - Required - 10 Years

****

JOB 3. Here's a job asking for a Project Manager who also will need some serious UI Designer skills, such as creating user stories and visual designs for those stories. This is that kind of job I used to take on in masquerade, using my project management-related designer skills to perform the PM role while happily focusing on the UI design tasks.

***

Title: Information Systems Project Manager
Company: Direct Client
Job ID#
Duration: 3-6 months (Contract to hire)
# of Openings: 1
Position Type: Contract to hire
Location:
Pay rate: only on $70/hr. on W2 or $74/hr. on 1099
No remote work.
Only US Citizens or Green card holder

Description/Comment: 3-6 months CTH Salary - 130K / yr

Northrop Grumman Information Systems sector is seeking an Information Systems Project Manager with a strong enterprise level Information Technology background to join our team of qualified, diverse individuals. This is an on-site position with no telecommute possibilities located in . This position is part of the Northrop Grumman's ETS Contract with the United States Postal Service, bringing top of the line professional technologists to their teams.

Roles and Responsibilities Include, but not limited to:
• Act as a communications liaison between teams and higher level management to include presentations, documentation and stories
• Work directly with management to define and develop metrics to measure key indicators including value, efficiency, effectiveness and progress to FY goals
• Compiles project data and reports on status of key projects to management and present metrics and stories into visual presentations
• Provide day-to-day project direction, ensuring quality standards, project

tracking, and analysis of assigned aspects of assigned project
• Transform technical content into elegant visual representations
• Positively influence peer organization and key stakeholders
• Run reports from Incident / Change Management systems (e.g. ServiceNow)
• Develop and present program status / metrics to senior US Government Client management
• Consult with management and customers regarding product feasibility and viability of project goals and include includes emerging technology and product supportability into design and implementation.
• Monitor project status and advertise the success of the team.
• Identify roadblocks and propose effective solutions
• Issue tracking & follow-up (drive deadlines)
• Make recommendations on staffing and quality improvement initiatives
• Ensure that multiple teams are in sync with the overall vision of the project.
• Serve as liaison between team members and customer management.
• Provide advanced support for the diagnosis and resolution of complex problems to ensure minimal disruption of mission-critical systems and applications.
• Leads teams and provides day-to-day direction for projects.

Basic Qualifications

• Bachelors' degree and a minimum of 9 years' of directly related work experience; or an additional 4 years' of experience may be substituted in lieu of the degree
• Previous experience and a complete understanding of technology management
• Past experience or knowledge of Agile methodology including sprints and stories
• Demonstrated excellence in planning, directing, and managing complex IT projects/operations
• Demonstrated leadership attributes (strategic planning, project planning, organizational management, etc.)
• Demonstrated strong knowledge of industry standard software development lifecycle
• Exceptional abilities with Power Point, communication and presentations.
• Excellent communication skills and ability to present to all levels of

management
• Ability to drive projects and provide day-to-day direction to team members.
• Experience with MS office products including advanced skills with Excel
• Must be able to obtain a Position of Public Trust Clearance –and must not have traveled outside the US for a combined total of 6 months or more in last 5 years.
• Must have resided in the US for the last 5 years

Preferred Qualifications:

• Bachelors degree and 15 years experience. Additional experience may be substituted for degree.
• MBA or advanced degree
• Agile Certifications
• PMP certificate
• ITIL Foundations certification

****

References

Dear Reader.

The references in this book are HTML links to online sources. The links are shown here. These links are also online for your ease of use. You can find these online links on my Amazon author page. To get to my Amazon author page, select my name on the Amazon page that advertises this book.

Eva

CH 1

1. Mainframe
   http://www-03.ibm.com/ibm/history/ibm100/us/en/icons/system360/transform/

CH 2

2. cardpunch
   https://m.youtube.com/watch?v=KG2M4ttzBnY

3. what is Script?
   https://en.m.wikipedia.org/wiki/SCRIPT_(markup)

4. philosophy
   https://en.m.wikipedia.org/wiki/Philosophy_of_artificial_intelligence

5. CAT
   https://en.m.wikipedia.org/wiki/Novation_CAT

6. Star Trek
   https://en.m.wikipedia.org/wiki/Star_Trek_(1971_video_game)

7. daisy wheel
   https://en.m.wikipedia.org/wiki/Daisy_wheel_printing

8. Kaypro II
http://www.classiccmp.org/dunfield/kaypro/index.htm

9. TSO on top of MVS
https://en.m.wikipedia.org/wiki/Time_Sharing_Option

10. WordStar
https://en.m.wikipedia.org/wiki/WordStar

CH 4

11. Agile Software
https://en.wikipedia.org/wiki/Agile_software_development

12. Waterfall
https://en.wikipedia.org/wiki/Waterfall_model

13. middleware
https://en.wikipedia.org/wiki/Middleware

14. UML
https://en.wikipedia.org/wiki/Unified_Modeling_Language

15. Use Case
https://en.wikipedia.org/wiki/Use_case

16. Dreamweaver
https://en.wikipedia.org/wiki/Adobe_Dreamweaver

17. Alistair Cockburn
https://en.wikipedia.org/wiki/Alistair_Cockburn

18. software startup companies
https://www.infoq.com/articles/what-do-we-know-about-software-development-in-startups/

19. haiku error messages
https://www.gnu.org/fun/jokes/error-haiku.en.html

20. Interface Hall of Shame
http://hallofshame.gp.co.at/shame.htm

CH 5

21. Readme file
https://en.wikipedia.org/wiki/README

22. *About Face*
https://www.amazon.com/About-Face-Essentials-Interaction-Design/dp/1118766571

23. Usability Professionals Association (UPA)
https://www.interaction-design.org/literature/publisher/usability-professionals-association

24. web spiders
https://en.wikipedia.org/wiki/Web_crawler

25. Here are some examples
https://www.cbinsights.com/research/dot-com-bubble-companies/

CH 8

26. Sir Jonathan Paul "Jony" Ive
https://en.wikipedia.org/wiki/Jony_Ive

CH 9

27. George Carlin's famous routine
https://www.google.com/search?q=george+carlin+routine+about+stuff&oq=george+carlin+routine+&aqs=chrome.1.69i57j0l5.6482j0j7&sourceid=chrome&ie=UTF-8

28. 92,000 public schools
https://www.statista.com/statistics/238307/number-of-us-elementary-and-secondary-schools/

29. TARP
https://en.wikipedia.org/wiki/Troubled_Asset_Relief_Program

30. Obamacare website
https://www.politifact.com/factchecks/2015/jul/14/donald-trump/donald-trump-says-5-billion-website-obamacare-neve/

31. 36% of our GDP (17.43 trillion dollars)
https://www.nato.int/nato_static_fl2014/assets/pdf/pdf_2019_11/20191129_pr-2019-123-en.pdf

https://www.google.com/search?q=US+gdp+in+2014&oq=US+gdp+in+2014&aqs=chrome.69i57j0l7.3981j0j4&sourceid=chrome&ie=UTF-8

32. Electronic Health Records
https://www.ncbi.nlm.nih.gov/pmc/articles/PMC5171496/

33. SUV today weighs at least 4,000 pounds
https://autoshopaccessories.com/how-much-does-a-car-weigh/#Midsize_SUV

AP 3

34. website of Tyner Blain
http://tynerblain.com/blog/2007/04/09/sample-use-case-example/

35. Creative Commons Attribution 4.0 International License
https://creativecommons.org/licenses/by/4.0/

About The Author

Eva Springstone

To My Readers:

You can reach me at:

evaspringstone@gmail.com

Warm regards,

Eva

Made in United States
North Haven, CT
04 November 2026

21749002R00146